# Electronic Circuit Guidebook
## Volume 1:
## Sensors

Written By
Joseph J. Carr

A Division of Howard W. Sams & Company
A Bell Atlantic Company
Indianapolis, IN

**PROMPT©** **Publications** is an imprint of Howard W. Sams & Company, A Bell Atlantic Company, 2647 Waterfront Parkway, E. Dr., Indianapolis, IN 46214-2041.

**International Standard Book Number: 0-7906-1098-1**
**Library of Congress Card Catalog Number: 97-65787**

*Acquisitions Editor*: Candace M. Hall
*Editors*: Natalie F. Harris, Loretta L. Leisure
*Assistant Editors*: Pat Brady
*Typesetting*: Charlene Flechsig, Leah Marckel
*Cover Design*: Suzanne Lincoln
*Graphics Conversion*: Debra Delk, Bill Skinner, Walt Striker, Terry Varvel
*Illustrations and Other Materials*: Courtesy of the Author

PRINTED IN THE UNITED STATES OF AMERICA

9 8 7 6 5 4 3 2 1

# Contents

# Contents

# Chapter 18

## Using Microcontrollers in the Instrument Front-End

# Chapter 19

## Sensor Resolution Improvement Techniques

# Chapter 20

## DC Power Supplies for Sensor-Based Circuits

# Preface

Electronic instrumentation is largely computer-based today. The use of the digital computer has greatly improved the design of electronic instruments, control systems and related devices (although not without introducing a few new problems...but's that the march of progress). A lot of texts are available that discuss electronic instrumentation and computer interfacing. A topic that is lacking, however, is the analog interface.

Most sensors are inherently analog in nature, so their outputs are not usable by the digital computer. Even if the sensor is supposedly a "digital output" design, it is likely that some inherently analog process is paired with an analog-to-digital converter. In this book you will find information about typical sensors, along with a large amount of information about analog sensor circuitry. Amplifier circuits are especially well covered.

A chapter is also provided on analog signal processing circuits such as differentiators and integrators. These functions are often done in software, but there are times when they must be done in the analog front-end.

This book is intentionally kept practical in outlook. To further the goals of the typical instrument designer or user, chapters on shielding of sensors and analog circuitry are provided, as well as specifics about eliminating 60 Hz ac power line interference from sensor circuits. Microcontrollers are an emerging technology that holds much promise for the sensor-based instrumentation designer or user. A chapter is provided that will introduce you to this fascinating class of devices.

The chapter on dc power supplies does not discuss the basics of dc power supplies, except in passing (and then briefly). The topics covered in this chapter are the distribution and control issues that you will face whether you build your own dc power supply or buy an off-the-shelf product.

Finally, there is a chapter on an exciting method for improving the resolution of sensors. A tremendous improvement is possible using this technology that is based on radar technology.

# Introduction to Sensor-Based Instrumentation

Sensors, transducers and electrodes are used in scientific, engineering and industrial instrument circuits to measure physical parameters by producing an output current or voltage signal that represents that parameter. For example, a thermocouple is a junction of two dissimilar metals that produces a voltage that is proportional to the applied temperature. Similarly, a piezoresistive strain gage produces an output voltage that is proportional to strain on a resistance element, which in turn is proportional to an applied displacement, force or pressure.

One thing that most sensors have in common is that they produce an analog voltage (V) or current (I) as the output. Depending on the specific nature of the sensor, there will be some correspondence between the applied stimulus being measured and the analog V or I output. Sometimes that correspondence is linear, in other cases it is logarithmic or some other nonlinear function. Regardless of the specifics for any given sensor, there will be an analog signal.

It is also true that most instruments and control systems used in industry, science, engineering and medicine today are based on a programmable digital computer of one sort or another. In some cases, that computer is a single-chip microcontroller. In other cases it is a desktop personal computer, or perhaps an embedded single-board computer. In still other cases the computer might be a large mini- or mainframe computer. What they all have in common, however, is a distinct inability to read the output of analog sensors, even though analog sensors dominate the industry.

A device or circuit called an analog-to-digital converter (A/D) produces a binary number or "word" that the computer can read in response to an analog input voltage. This would seem to be the end of it, but that is not the case. There are other things that the analog subsystem ahead of the computer-based instrument must do prior to sending the signal to the computer.

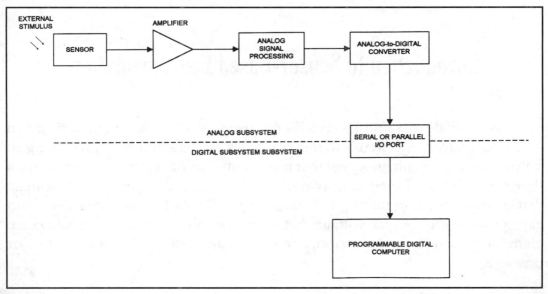

*Figure 1-1. A typical analog subsystem.*

Figure 1-1 shows a block diagram of a typical analog subsystem. A sensor is impinged with some external stimulus producing a voltage or current output signal. This signal usually must be amplified to a higher level in order to get it out of the noise or to fill up the range of the A/D converter. After all, a costly 16-bit A/D converter is wasted if the maximum signal voltage only fills up 8, 10 or 12 bits.

There might also be some analog signal processing circuits. Most A/D converters ought to have a filter ahead of them to prevent noise problems or aliasing. There is also a need on occasion for a logarithmic amplifier to improve dynamic range, or an integrator to provide a time average or low-pass filtering action. There are a number of different analog signal processing circuits needed in some systems.

Finally, after a certain amount of signal conditioning, one gets to the A/D converter, which is very much a part of the analog subsystem. I have drawn the boundary of the analog subsystem at the data transmission interface (which could be either a serial or parallel port), although a good argument could also be made for drawing the boundary in the middle of the A/D converter...it being analog on one side and digital on the other. I drew the dotted line the way it is because many analog subsystems are equipped with a parallel or serial port function in order to allow it to communicate with the computer. Rarely is the raw data output of the A/D converter satisfactory for sending directly to the computer.

Some people scoff at the idea that the analog subsystem is important, but I suspect those are people who never had to build a sensor-based data acquisition system. Sometimes you will off-load small computers of tasks that are easily handled in analog circuitry. Other times, especially where small embedded computers or microcontrollers are concerned, either system resources are insufficient or cost is prohibitive. The analog solution becomes attractive. Although I would agree that (system resources permitting) it is generally better to implement functions in the computer software, it is not universally the case.

There are even some jobs where the analog subsystem is either essential or does better than computers. For example, some optoelectronic sensors produce signals that, at low light levels, are barely above the noise floor. When amplified to a level where the A/D converter could function, the noise becomes problematic; it tends to fill the dynamic range "head room" of the A/D circuit. A heavy dose of filtering or integration works wonders. Another problem is when the desired signal is riding on a very large pedestal such as a DC offset. That occurs in measuring light variation in relatively bright ambient situations, small fluctuations in the Earth's magnetic field, or even the human electrocardiograph (ECG). In the latter case, a 1 µV heart signal rides on a 1-2 volt pedestal created by the battery-like action of the electrode against the patient's skin. That 1000:1 ratio can sometimes be handled in the A/D converter by using something on the order of a 22-bit or 24-bit A/D and then stripping off the ten least significant bits in order to get the 10-bit resolution that is needed (some prefer 12-bit).

This book explores some common sensor types (by way of example), the analog circuitry that supports the sensors, and some of the issues that arise when building the analog subsystem. It needs to be taken seriously by instrument and control system designers because it is, after all, the interface between the sensor and the computer.

# Chapter 2

# The Nature of Electronic Signals

The nature of electrical signals, and their relationship to noise and interfering signals, determines appropriate instrument design all the way from the system level down to the component selection level. In this chapter we will take a look at electronic signals.

Electronic signals can be categorized several ways, but one of the most fundamental is according to time domain behavior (the other major category is frequency domain). It is therefore useful to consider voltage (V) and current (I) signals of the form V = f(t) or I = f(t), where f represents frequency and t equals time period. The time domain classes of signals include: static, quasistatic, periodic, repetitive, transient, and quasitransient. Each of these categories have certain properties that can profoundly influence appropriate design decisions.

## Static and Quasistatic Signals

A static signal (Figure 2-1a) is, by definition, unchanging over a very long period of time ($T_{long}$ in Figures 2-1a and 2-1b). Such a signal is essentially a dc level, so must be processed in low-drift dc amplifier circuits. The term quasistatic means "nearly unchanging," so a quasistatic signal (Figure 2-1b) refers to a signal that changes so slowly over long times that it possesses characteristics more like those of static signals than dynamic (i.e. rapidly changing) signals.

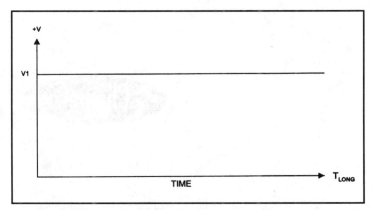

*Figure 2-1a. Constant dc (static) signal.*

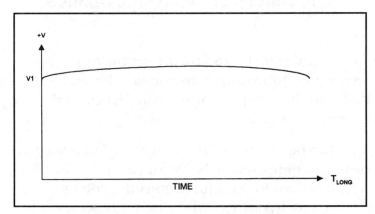

*Figure 2-1b.  Quasistatic slow changing signal.*

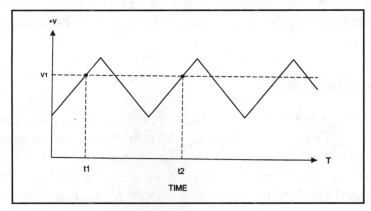

*Figure 2-1c.  Periodic signal.*

## Periodic Signals

A periodic signal (Figure 2-1c) is one that repeats itself on a regular - that is, periodic - basis. Examples of periodic signals include sine waves, square waves, sawtooth waves, triangle waves, and so forth. The nature of the periodic waveform is such that each waveform is identical at like points along the time line. In other words, if you advance along the time line by exactly one period (T), then the voltage, polarity and direction of change of the waveform will be repeated. That is, for a voltage waveform of period T, $V(t) = V(t + T)$.

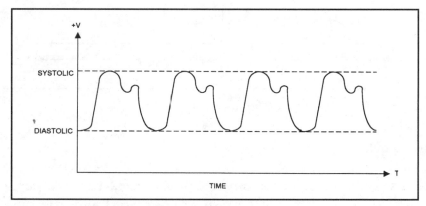

*Figure 2-1d. Repetitive signal.*

## Repetitive Signals

A repetitive signal (Figure 2-1d) is nearly periodic in nature, so bears some similarity to the periodic waveform. The principal difference between repetitive and periodic signals is seen by comparing the signal at $f(t)$ and $f(t + T)$. These points might not be identical in repetitive signals, but are identical in periodic signals. The repetitive signal might contain either transient or stable features that vary from period to period. An example is the human arterial blood pressure waveform (Figure 2-1d). While the waveform tends to vary from a minima (diastolic) to a maxima (systolic) in a nearly periodic manner, there are both normal and pathological anomalies from one cycle to another. For example, the amplitudes of the maxima and minima, and the repetition rate (i.e. heart rate) tend to normally vary quite a bit in healthy humans. In addition, events such as premature ventricular contractions (PVC) are anomalies that may or may not be pathological (PVCs tend to be purely transient events superimposed on the repetitive signal). Thus, the repetitive signal may bear characteristics of both transient and periodic signals.

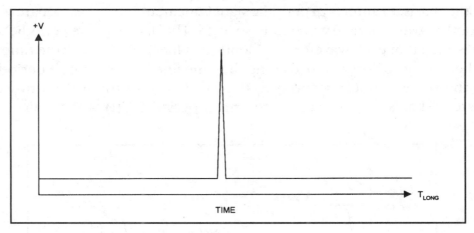

*Figure 2-1e.   Transient signal.*

## Transient Signals

A transient signal is either a onetime event (Figure 2-1e), or a periodic event in which the event duration is very short compared with the period of the waveform (such signals are sometimes called quasitransient). In terms of Figure 2-1f, the latter definition means that T1 << T2. These signals can be treated as if they are transients.

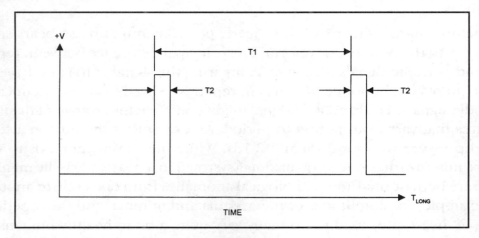

*Figure 2-1f.   Quasitransient signal.*

The characteristic of these various types of signals often drive the design of the instrumentation system. An important consideration of all dynamic signals (whether periodic or otherwise) is the frequency response required to faithfully reproduce the signal. For that reason we will turn our attention to the frequency domain characteristics of signals.

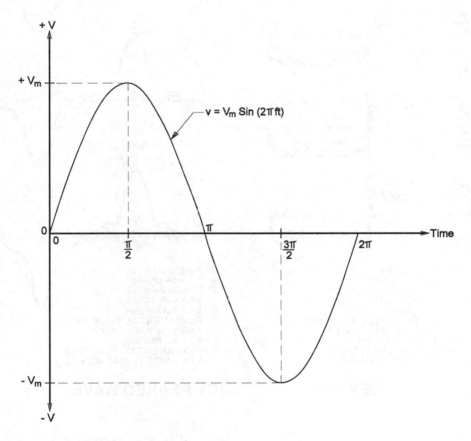

*Figure 2-2. Sine wave signal.*

## Fourier Series

All continuous periodic signals can be represented by a fundamental frequency sine wave, and a collection of sine and/or cosine harmonics of that fundamental sine wave, that are summed together linearly. These frequencies comprise the Fourier series of the waveform. The elementary sine wave (Figure 2-2) is described by:

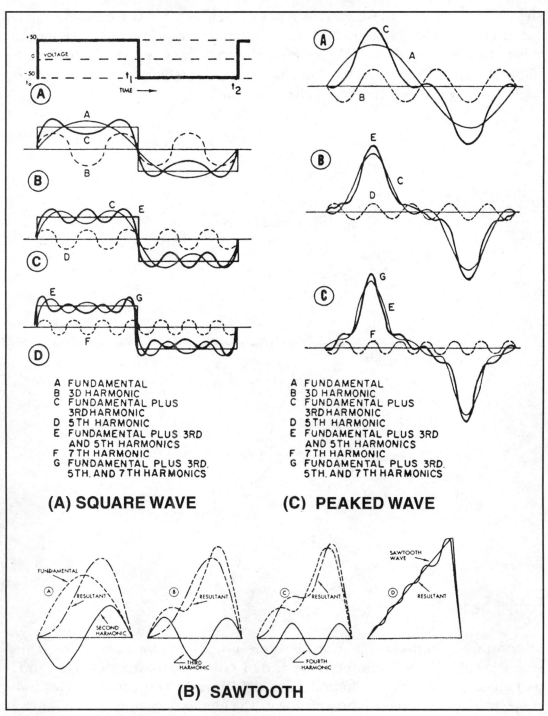

A FUNDAMENTAL
B 3D HARMONIC
C FUNDAMENTAL PLUS
  3RD HARMONIC
D 5TH HARMONIC
E FUNDAMENTAL PLUS 3RD
  AND 5TH HARMONICS
F 7TH HARMONIC
G FUNDAMENTAL PLUS 3RD,
  5TH, AND 7TH HARMONICS

**(A) SQUARE WAVE**

A FUNDAMENTAL
B 3D HARMONIC
C FUNDAMENTAL PLUS
  3RD HARMONIC
D 5TH HARMONIC
E FUNDAMENTAL PLUS 3RD
  AND 5TH HARMONICS
F 7TH HARMONIC
G FUNDAMENTAL PLUS 3RD,
  5TH, AND 7TH HARMONICS

**(C) PEAKED WAVE**

**(B) SAWTOOTH**

*Figure 2-3.   Signal composition.*

$$v = V_m \sin(2\omega t) \qquad \qquad eq. \; (2\text{-}1)$$

Where:

v is the instantaneous amplitude of the sine wave.

$V_m$ is the peak amplitude of the sine wave.

$\omega$ is the angular frequency ($2\pi f$) of the sine wave.

t is the time in seconds.

The period of the sine wave is the time between reoccurrence of identical events, or $T = 2\pi/\omega = 1/f$ (where $f$ is the frequency in cycles per second).

The Fourier series that makes up a waveform can be found if a given waveform is decomposed into its constituent frequencies either by a bank of frequency selective filters, or a digital signal processing algorithm called the Fast Fourier Transform (FFT). The Fourier series can also be used to construct a waveform from the ground up. Figure 2-3 shows square wave (Figure 2-3a), sawtooth wave (Figure 2-3b) and peaked wave (Figure 2-3c) signals constructed from fundamental sine waves and their harmonic sine and cosine functions.

The Fourier series for any waveform can be expressed in the form:

$$f(t) = \frac{a_o}{2} + \int_{n=1}^{\infty} a_n \, Cos(n\omega t) + b_n \, Sin(n\omega t) \, dt$$

$$eq. \; (2\text{-}2)$$

Where:

$a_n$ and $b_n$ represents the amplitudes of the harmonics (see below).

n is an integer.

Other terms are as previously defined.

The amplitude coefficients ($a_n$ and $b_n$) are expressed by:

$$a_n = \frac{2}{T} \int_0^t f(t) \cos(n\omega t)\, dt \qquad \text{eq. (2-3)}$$

and,

$$b_n = \frac{2}{T} \int_0^t f(t) \sin(n\omega t)\, dt \qquad \text{eq. (2-4)}$$

The amplitude terms are nonzero at the specific frequencies determined by the Fourier series. Because only certain frequencies, determined by integer n, are allowable, the spectrum of the periodic signal is said to be discrete.

The term $a_o/2$ in the Fourier series expression (Equation 2-2) is the average value of f(t) over one complete cycle (one period) of the waveform. In practical terms, it is also the dc component of the waveform. When the waveform possesses half wave symmetry (i.e. the peak amplitude above zero is equal to the peak amplitude below zero at every point in t, or $+V_m = |{-V_m}|$), there is no dc component, so $a_o = 0$.

An alternative Fourier series expression replaces the $a_n \cos(n\omega t) + b_n \sin(n\omega t)$ with an equivalent expression of another form:

$$f(t) = \frac{2}{T} \int_0^\infty c_n(n\omega t - \phi_n)\, dt \qquad \text{eq. (2-5)}$$

Where:

$$c_n = \sqrt{(a_n)^2 + (b_n)^2}$$

$$\phi_n = \arctan(a_n / b_n)$$

All other terms are as previously defined.

One can infer certain things about the harmonic content of a waveform by examination of its symmetries. One would conclude from the above equations that the harmonics extend to infinity on all waveforms. Clearly, in practical systems, a much less than infinite bandwidth is found, so some of those harmonics will be removed by the normal action of the electronic circuits. Also, it is sometimes found that higher harmonics might not be truly significant, so can be ignored. As *n* becomes larger, the amplitude coefficients $a_n$ and $b_n$ tend to become smaller. At some point, the amplitude coefficients are reduced sufficiently that their contribution to the shape of the wave is either negligible for the practical purpose at hand, or are totally unobservable in practical terms. The value of n at which this occurs depends partially on the rise time of the waveform. Rise time is usually defined as the time required for the waveform to rise from 10 percent to 90 percent of its final amplitude. Let's consider a practical example from biomedical instrumentation.

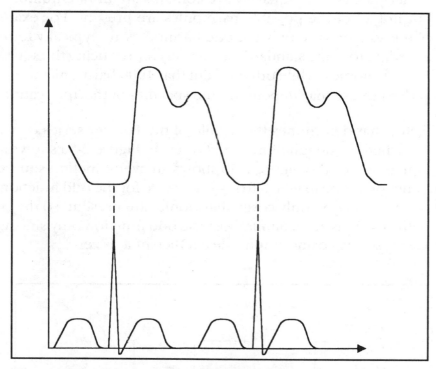

*Figure 2-4. Human arterial pressure and ECG signals.*

Figure 2-4 shows the human arterial pressure waveform superimposed on the same time line as the electrocardiograph (ECG) waveform. These waveforms are time correlated because they both represent different views of the same physical event, i.e. the beating of the human heart. Suppose that the heart rate is 72 beats per minute (BPM), or 1.2 Hz. The pressure waveform has a slower rise time than the ECG waveform, so contains a smaller number of harmonics. The pressure waveform can be accurately reproduced with about 25 harmonics (e.g. approximately 30 Hz bandwidth), while the ECG waveform requires 70 to 80 harmonics for faithful reproduction (e.g., a bandwidth of about 100 Hz). In order to adequately process these two waveforms the instrument must have upper -3 dB frequency responses of 30 Hz and 100 Hz for the pressure and ECG channels, respectively. Because both pressure and ECG waveforms have significantly rounded features, the lower -3 dB frequency response (a function of subharmonic content) must be 0.05 Hz.

The square wave represents another case altogether because it has a very fast rise time. Theoretically, the square wave contains an infinite number of harmonics, but not all of the possible harmonics are present. For example, in the case of the square wave only the odd harmonics are typically found (e.g. 3, 5, 7). According to some standards, accurately reproducing the square wave requires 100 harmonics, while others claim that 1000 harmonics are needed. Which standard to use may depend on the specifics of the application.

Another factor that determines the profile of the Fourier series of a specific waveform is whether the function is odd or even. Figure 2-5a shows an odd-function square wave, and Figure 2-5b shows an even-function square wave. The even function is one in which $f(t) = f(-t)$, while for the odd function $-f(t) = f(-t)$. In the even function only cosine harmonics are present, so the sine amplitude coefficient $b_n$ is zero. Similarly, in the odd function only sine harmonics are present, so the cosine amplitude coefficient $a_n$ is zero.

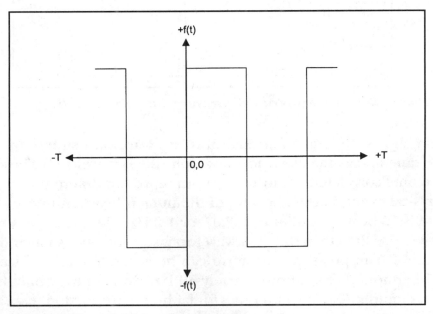

*Figure 2-5a. Odd-function square wave.*

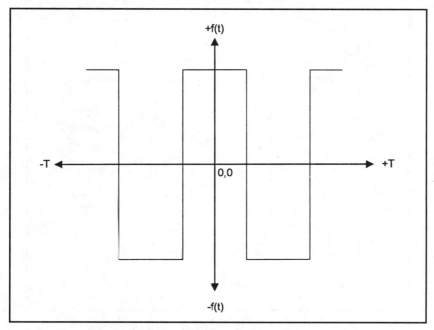

*Figure 2-5b. Even-function square wave.*

## Waveform Symmetry

Both symmetry and asymmetry can occur in several ways in a waveform (Figure 2-6), and those factors can affect the nature of the Fourier series of the waveform. In Figure 2-6a we see the case of a waveform with a dc component. Or, in terms of the Fourier series equation, the term ao is nonzero. The dc component represents a case of asymmetry in a signal. This offset can seriously affect instrumentation electronic circuits that are dc-coupled, and thereby result in serious artifact.

Two different forms of symmetry are shown in Figure 2-6b. Zero-axis symmetry occurs when, on a point-for-point basis, the waveshape and amplitude above the zero baseline is equal to the amplitude below the baseline (or $|+V_m|$ = $|-V_m|$). When a waveform possesses zero-axis symmetry it will usually not contain even harmonics, only odd harmonics are present. This situation is found in square waves, for example (Figure 2-7a). Zero-axis symmetry is not found only in sine and square waves, however, as the sawtooth waveforms in Figure 2-6c demonstrate.

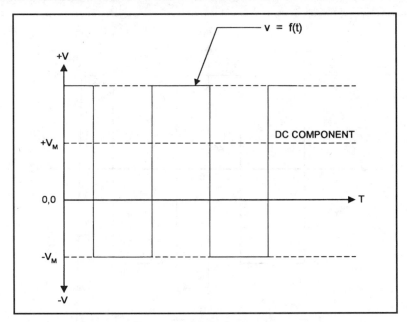

*Figure 2-6a.  Square wave with dc component.*

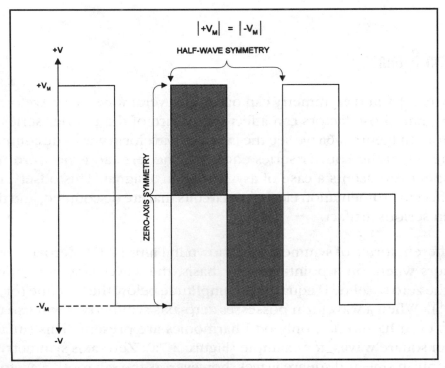

*Figure 2-6b.  Half-wave and zero-axis symmetry.*

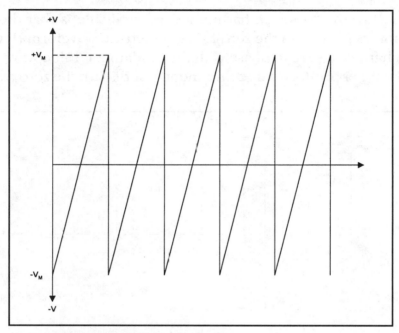

Figure 2-6c.  Sawtooth wave form.

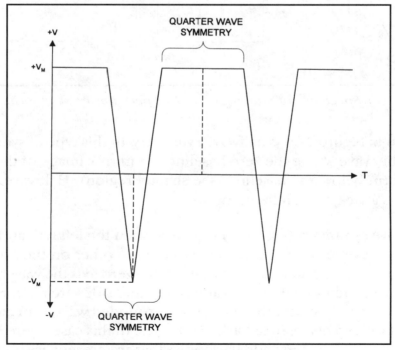

Figure 2-6d.  Quarter-wave symmetry.

An exception to the "no even harmonics" general rule is that there will be even harmonics present in the zero-axis symmetrical waveform (Figure 2-7b) if the even harmonics are in phase with the fundamental sine wave. This condition will neither produce a dc component, nor disturb the zero-axis symmetry.

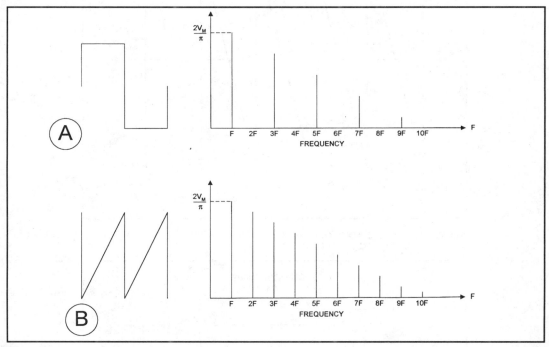

*Figure 2-7.  a) Spectrum of square wave; b) Spectrum of sawtooth.*

Also shown in Figure 2-6b is halfwave symmetry. In this type of symmetry the shape of the wave above the zero baseline is a mirror image of the shape of the waveform below the baseline (see shaded region). Halfwave symmetry also implies a lack of even harmonics.

Quarter-wave symmetry (Figure 2-6d) exists when the left half and right half sides of the waveforms are mirror images of each other on the same side of the zero-axis. Note in Figure 2-6d, that above the zero-axis the waveform is like a square wave, and indeed the left- and right-hand sides are mirror images of each other. Similarly, below the zero-axis the rounded waveform has a mirror image relationship between left and right sides. In this case, there is a full set of even harmonics, and any odd harmonics that are present are in phase with the fundamental sine wave.

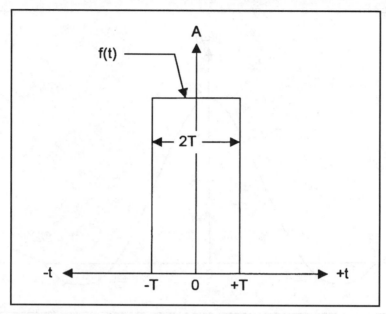

*Figure 2-8a.   f(t) vs. t spectrum.*

## Transient Signals

A transient signal is an event that occurs either once only, or occurs randomly over a long period of time, or is periodic but has a very short duration compared with its period (i.e. it is a very short duty cycle event). Many pulse signals fit the latter criterion even though mathematically they are actually periodic.

Transient signals are not represented properly by the Fourier series, but can nonetheless be represented by sine waves in a spectrum. The difference is that the spectrum of the transient signal is continuous rather than discrete (as in a periodic signal). Consider a transient signal of period 2T, such as Figure 2-8a. The spectral density, g($\omega$), is:

$$g(\omega) = \int_{-\infty}^{+\infty} f(t) e^{j\omega t} \, dt \qquad\qquad \text{eq. (2-6)}$$

Given a spectral density, however, the original waveform can be reconstructed from:

$$f(t) = \frac{1}{2\pi} \int_{-\infty}^{+\infty} g(\omega) e^{j\omega t} \, d\omega \qquad\qquad \text{eq. (2-7)}$$

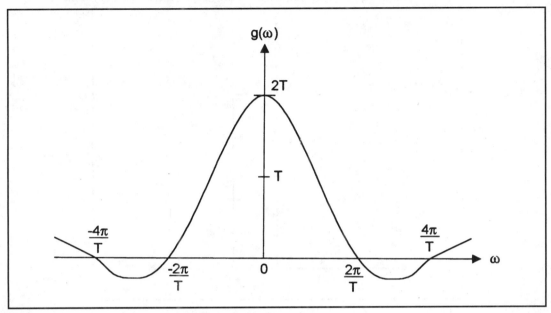

*Figure 2-8b. Sin x/x spectrum.*

The shape of the spectral density region is shown in Figure 2-8b. Note that the negative frequencies are a product of the mathematics, and do not have physical reality. The shape of Figure 2-8b is expressed by:

$$\bullet \quad g(\omega) \ = \ \frac{\sin\ \omega t}{\omega t} \qquad\qquad eq.\ (2\text{-}8)$$

The general form sin x/x is used also for repetitive pulse signals as well as the transient form shown in Figure 2-8b.

## Sampled Signals

The digital computer is incapable of accepting analog input signals, but rather requires a digitized representation of that signal. The analog-to-digital (A/D) converter (see Chapter 17) will convert an input voltage (or current) to a representative binary word. If the A/D converter is either clocked or allowed to run asynchronously according to its own clock, then it will take a continuous string of samples of the signal as a function of time. When combined, these signals represent the original analog signal in binary form.

But the sampled signal is not exactly the same as the original signal, and some effort must be expended to ensure that the representation is as good as possible. The waveform in Figure 2-9a is a continuous voltage function of time, V(t); in this case a triangle waveform is seen. If the signal is sampled by another signal, P(t), with frequency $F_s$ and sampling period $T = 1/F_s$, as shown in Figure 2-9b, and then later reconstructed, the waveform may look something like Figure 2-9c. While this may be sufficiently representative of the waveform for many purposes, it would be reconstructed with greater fidelity if the sampling frequency ($F_s$) is increased.

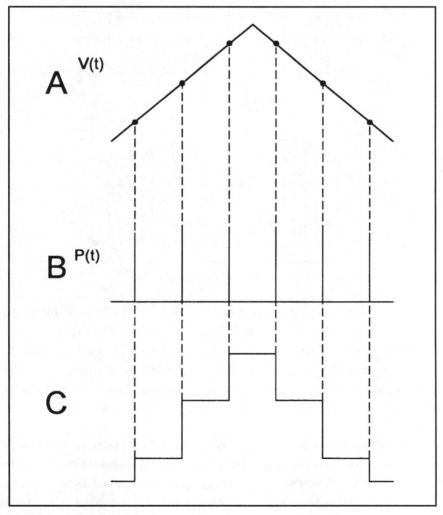

*Figure 2-9. Undersampled signal.*

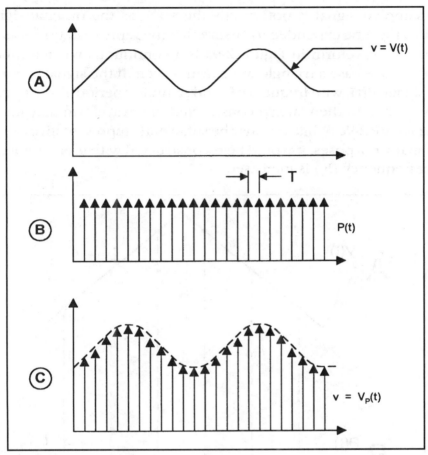

*Figure 2-10.    Properly sampled signal.*

Figure 2-10 shows another case in which a sine wave, V(t) in Figure 2-10a, is sampled by a pulse signal, p(t) in Figure 2-10b. The sampling signal, p(t), consists of a train of equally spaced narrow pulses spaced in time by T. The sampling frequency $F_s$ equals 1/T. The resultant is shown in Figure 2-10c, and is another pulsed signal in which the amplitudes of the pulses represent a sampled version of the original sine wave signal.

The sampling rate, $F_s$, must by Nyquist's theorem be twice the maximum frequency ($F_m$) in the Fourier spectrum of the applied analog signal, V(t). In order to reconstruct the original signal after sampling, it is necessary to pass the sampled waveform through a low-pass filter that limits the bandpass to $F_s$.

The sampling process is analogous to a form of amplitude modulation (AM), in which V(t) is the modulating signal, with spectrum from dc to $F_m$, and p(t) is the carrier frequency. The resultant spectrum is shown partially in Figure 2-11, and resembles the double sideband with carrier AM spectrum. The spectrum of the modulating signal appears as "sidebands" around the "carrier" frequency, shown here as $F_o$. The actual spectrum is a bit more complex, as shown in Figure 2-12. Like an unfiltered AM radio transmitter, the same spectral information appears not only around the fundamental frequency $(F_s)$ of the carrier (shown at zero in Figure 2-12), but also at the harmonics and subharmonics spaced at intervals of $F_s$ up and down the spectrum.

Providing that the sampling frequency $F_s > 2F_m$, the original signal is recoverable from the sampled version by passing it through a low-pass filter with a cut-off frequency $F_c$, set to pass only the spectrum of the analog signal - but not the sampling frequency. This phenomenon is shown with the dotted line in Figure 2-12.

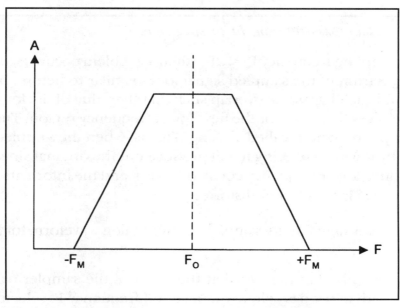

*Figure 2-11. Amplitude vs frequency spectrum.*

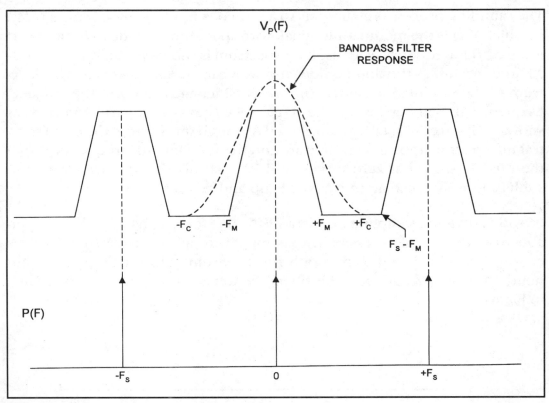

*Figure 2-12.   Bandpass filtering fit to spectrum.*

When the sampling frequency $F_s < 2F_m$, then a problem occurs (see Figure 2-13). The spectrum of the sampled signal looks similar to before, but the regions around each harmonic overlap such that the value of $-F_m$ for one spectral region is less than $+F_m$ for the next lower frequency region. This overlap results in a phenomenon called aliasing. That is, when the sampled signal is recovered by low-pass filtering it will produce not the original sine wave frequency $F_o$ but a lower frequency equal to $(F_s - F_o)$, and the information carried in the waveform is thus lost or distorted.

The solution, for high fidelity sampling of the analog waveform for input to a computer, is to:

    1. Bandwidth limit the signal at the input of the sampler or A/D converter with a low-pass filter with a cut-off frequency $F_c$ selected to pass only the maximum frequency in the waveform ($F_m$) and not the sampling frequency ($F_s$).

2. Set the sampling frequency $F_s$ at least twice the maximum frequency in the applied waveform's Fourier spectrum, i.e. $F_s > 2F_m$.

Experience has shown that some users will not accept a reconstructed sampled waveform if the sample rate is $2F_m$. For example, medical electrocardiograph (ECG) waveforms, in which $F_m = 100$ Hz, tend to look "blocky" when sampled at 200 Hz and then reconstructed. The user acceptance was much better when the waveform was sampled at 500 Hz, or $5F_m$. While that rate was expensive to accommodate in an eight-bit A/D converter at one time, it is now very low cost and should be used...Nyquist notwithstanding.

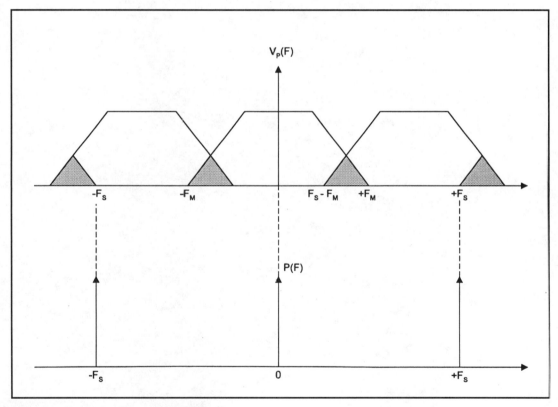

*Figure 2-13. Aliasing phenomenon.*

# Chapter 3

# Measurement, Sensors and Instruments

Sensors, transducers and electrodes are used in measurement and data acquisitions systems of various types. They may be the "machine perception organ" for a robot or servocontrol system. Or they may be part of an instrument that measures some physical parameter or another, and then display the result. For example, a thermistor (temperature sensor) could either be part of a servocontrol system in a factory process, or a digital thermometer for measuring the temperature in your mouth. Regardless of the specific sensor, the specific application, configuration or implementation, however, the sensor is part of a larger overall class of systems that measure.

The reason that the measurement is made is to fulfill one of several goals: obtain information about the physical process, record trends, control some process, or correlate behavior with other parameters in order to obtain insight into their relationship. When combined with each other and made to work properly, all of the elements of the measuring system comprise a data acquisition system.

According to E. E. Herceg, a measurement is an act that is designed to "derive quantitative information about..." some physical phenomenon "...by comparison to a reference..." or standard. The physical quantity being measured is called the measurand. The material starts out with a discussion of the various categories of measurement (e.g. direct, indirect, null, etc.). Also discussed are issues such as precision and accuracy (often erroneously confused), resolution, validity and reliability of a measurement. There is also a discussion of measurement error, and how to avoid some of the most serious and most common errors.

## Categories of Measurement

There are three general categories of measurement: direct, indirect and null. Electronic instruments are available based on all three categories.

Direct measurements are made by holding the measurand up to some calibrated standard and comparing the two. A good example is the meter stick ruler used to cut a piece of lumber to the correct length. You know that the wood must be cut to a length of 24 cm, so you hold a meter stick (the standard or reference) up to the uncut piece of lumber (Figure 3-1), set the "0 cm" point at one end, and make a mark on the cable adjacent to the "24" mark on the meter stick, and then make your cut at the appropriate point.

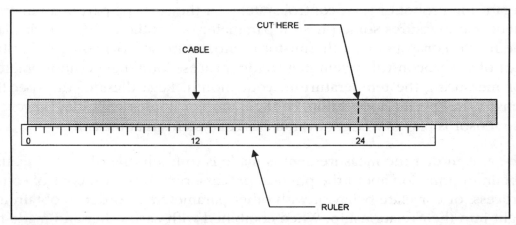

*Figure 3-1. Direct measurement system against a standard.*

Indirect measurements are made by measuring something other than the actual measurand. Although frequently considered "second best" from the perspective of measurement accuracy, indirect methods are often used when direct measurements are either difficult or dangerous. For example, one might measure the temperature of a point on the wall of a furnace that is melting metal, knowing that it is related to the interior temperature by a certain factor.

There was once a minicomputer manufacturer who used an indirect temperature measurement to ease the job of the service technicians. The equipment was one of the first commercial (i.e. non-research) computer-based systems for monitoring coronary care unit (CCU) patients' analog ECG and pressure waveforms. The manufacturer created a small hole at the top of the rack mounted cabinet where the temperature would be <39 °C when the temperature on the electronic circuit boards deep inside the cabinet was within specification. Although they were interested in the temperature at the board level, they actually make a measurement that correlates to the desired measurement.

The system manufacturer specified this method for two reasons: a) the measurement point was easily available (while the boards were not), and thus did not require any disassembly; and b) the service technician could use an ordinary medical fever thermometer (30 °C to 42 °C) as the measurement instrument ...which was easily available wherever the computer was installed. No special laboratory thermometers were needed.

Perhaps the most common example of an indirect measurement is the human blood pressure as ordinarily measured in doctor's offices. It is measured by measuring the pressure in an occluding cuff placed around the arm; a process called sphygmomanometry. Research around 1905 by Nicolai Korotkoff showed that the cuff pressures at two easily heard events (onset and cessation of "Korotkoff sounds") are correlated to the systolic ($P_s$) and diastolic ($P_d$) arterial blood pressures. Direct blood pressure measurement may be more accurate, but is dangerous because it is an evasive surgical procedure. This method does not actually measure the blood pressure, but rather it measures cuff pressure. By correlating cuff pressure with the onset and cessation of certain standard events called "Korotkoff sounds", a measure of blood pressure is possible.

Null measurements are made by comparing a calibrated source to an unknown measurand, and then adjusting either one or the other until the difference between them is zero. An electrical potentiometer is such an instrument; it is an adjustable calibrated voltage source and a comparison meter (zero-center galvanometer). The reference voltage from the potentiometer is applied (Figure 3-2) to one side of the zero-center galvanometer (or one input of a difference measuring voltmeter), and the unknown is applied to the other side of the galvanometer (or remaining input of the differential voltmeter). The output of the potentiometer is adjusted until the meter reads zero difference. The setting of the potentiometer under the nulled condition is the same as the unknown measurand voltage.

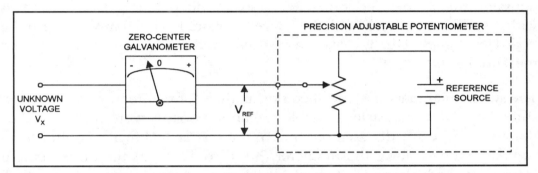

*Figure 3-2. Comparison or transfer measurement.*

## Quality Factors in Making Measurements

The "goodness" of measurements involves several concepts that must be understood. Some of the more significant of these are: error, validity, reliability, repeatability, accuracy, precision, and resolution.

Error. In all measurements there is a certain degree of error present. The word "error" in this context refers to normal random variation, and in no way means "mistakes." In short order we will discuss error in greater depth.

If measurements are made repeatedly on the same parameter (which is truly unchanging), or if different instruments or instrument operators are used to make successive measurements, it will be found that the measurements tend to cluster around a central value ($X_o$ in Figure 3-3). In most cases, it is assumed that $X_o$ is the true value, but if there is substantial inherent error in the measurement process, then it may deviate from the true value ($X_i$) by a certain amount ($\Delta X$) - which is the error term. The assumption that the central value of a series of measurements is the true value is only valid when the error term is small. As $\Delta X \rightarrow 0$, $X_o \rightarrow X_i$.

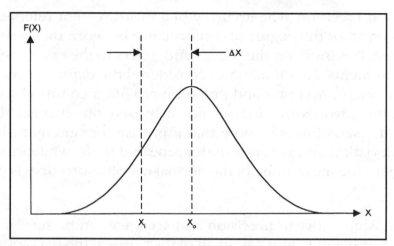

*Figure 3-3. Error ΔX is an indication of how far the actual value is from the measured value.*

Validity. The validity of a measurement is a statement of how well the instrument actually measures what it purports to measure. An electronic blood pressure sensor may actually be measuring the deflection of a thin metallic diaphragm of known area, which is in turn measured by the strain applied to a strain gauge element cemented to the diaphragm. What determines the validity of a sensor measurement is the extent to which the measurement of the deflection of that diaphragm relates to applied pressure, and over what range or under what conditions. Many measurement devices exist where the output readings are only meaningful under certain specified conditions or over a specified range.

Reliability and Repeatability. The reliability of the measurement is a statement of its consistency when discerning the values of the measurand on different trials, when the measurand may take on very different values. In the case of the pressure sensor discussed above, a deformation of a diaphragm may change its characteristics sufficient to alter future measurements of the same pressure value.

Related to reliability is the idea of repeatability, which refers to the ability of the instrument to return the same value when repeatedly exposed to the exact same stimulant. Neither reliability nor repeatability are the same as accuracy, for a measurement may be both "reliable" and "repeatable" while being quite wrong.

Accuracy and Precision. The accuracy of a measurement refers to the freedom from error, or the degree of conformance between the measurand and the standard. Precision, on the other hand, refers to the exactness of successive measurements, also sometimes considered the degree of refinement of the measurement. Accuracy and precision are often confused with one another, and these words are often erroneously used interchangeably. One way of stating the situation is to note that a precise measurement has a small standard deviation and variance under repeated trials, while in an accurate measurement the mean value of the normal distribution curve is close to the true value.

The relationship between precision and accuracy can be seen in the target shooting example of Figure 3-4. In all of these cases, the data form a normal distribution curve when repeatedly performed over a large number of iterations of the measurement. Four targets are shown in a precision-vs-accuracy matrix. Target 3-4a has good accuracy because the shots are clustered on the bullseye. It also has good precision, as seen by the fact that the cluster has a small dispersion, i.e. it is a "tight group" as target shooters say. The target at 3-4b has good precision (small dispersion, good clustering), but the cluster is off center, high and to the left. The target a 3-4c has good accuracy because the cluster is centered, but the bullet holes are all over the paper, which indicates a lack of precision. The target in 3-4d lacks both accuracy and precision.

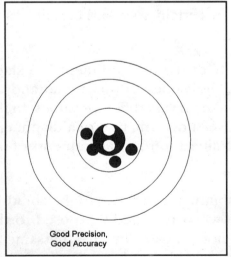

Figure 3-4a.  Good accuracy, good precision.

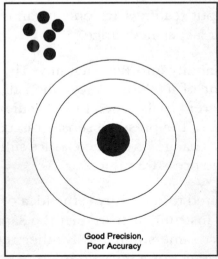

Figure 3-4b.  Poor accuracy, good precision.

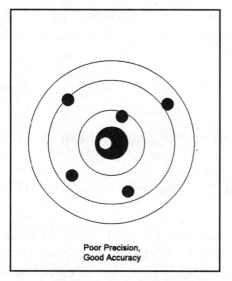

Figure 3-4c.  Good accuracy, poor precision.

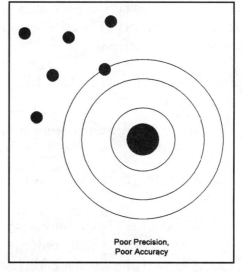

Figure 3-4d.  Poor accuracy, poor precision.

Target shooting is a good analogy for measurement processes and how to solve problems with them. Good shooting instructors know that it's better to work on precision first, i.e. getting the cluster smaller (called "grouping" in shooting). That's analogous to reducing the random or inherent variation in a measurement process. Some of the clustering is due to the mechanics of the gun, but most of it happens to be due to the shooter. Once the shooter is consistently shooting tight clusters, then it's time to worry about moving the impact point (i.e. moving the average). How is this done? Not by adjusting the shooter, but adjusting the process (i.e. the gun). The gun that shot the target at B can be brought into good working order by adjusting the movable sights about two clicks to the right and two clicks down.

A mistake often made by novice shooters, and poor engineers, is to adjust the sights with too few shots on the paper. Enough shots (data points) must be collected to truly see the clustering before any meaningful change can be made. I have observed shooters firing two shots and then adjusting the sights; firing two more shots and adjust; fire two more shots and adjust, etc. They wonder why the gun never seems to come into regulation. They blame the gun, the ammunition and the lighting on the range, but never their methods.

The standard deviation (which is the square root of variance) of the measurement is a good indication of its precision, which also means the inherent error in the measurement.

There are several tactics that help reduce the effects of error on practical measurements:

1. Make the measurement a large number of times, and then average the results.

2. Make the measurement several times using different instruments, if feasible.

3. When using instruments such as rulers or analog meters, try making the successive measurements on different parts of the scale. For example, on rulers and analog meter dials the distance between tick marks is not really constant because of manufacturing error. The same is also true of electrical meter scales. Measure lengths using different points on the scale as the zero reference point (e.g. on a meter stick use 2, 12, 20, and 30 cm as the zero point), and then average the results. By taking the measurements from different sections of the scale, both individual errors and biases that accumulate will be averaged to a lower overall error.

Resolution. This term refers to the degree which the measurand can be broken into identifiable adjacent parts. An example can be seen on the standard television test pattern broadcast by some stations in the early morning hours between "broadcast days." Various features on the test pattern will include parallel vertical or horizontal lines of different densities. One patch may be 100 lines per inch, another 200 lines per inch, and so forth up the scale. The resolution of the video system is the maximum density at which it is still possible to see adjacent lines with space between them. For any system, there is a limit above which the lines are blurred into a single entity.

In a digital electronic measuring instrument, the resolution is set by the number of bits used in the data word. Digital instruments use the binary (base-2) numbers system in which the only two permissible digits are "0" and "1." The binary "word" is a binary number representing a quantity. For example, binary "0001" represents decimal 1, while binary "1001" represents decimal

9. An 8-bit data word, the standard for many small embedded computers, can take on values from $00000000_2$ to $11111111_2$, so can break the range into $2^8$ (256) distinct values, or $2^8-1$ (255) different segments. The resolution of that system depends on the value of the measured parameter change that must occur in order to change the least significant bit in the data word. For example, if 8-bits are used to represent a voltage range of 0 to 10 volts, then the resolution is (10 - 0)/255, or 0.039 V (e.g. 39 µV) per bit. This resolution is often specified as the "1-LSB" resolution.    mV

## Measurement Errors

No measurement is perfect, and measurement apparatus are never ideal, so there will always be some error (not mistakes) in all forms of measurement. An error is a deviation between the actual value of a measurand and the indicated value produced by the sensor or instrument used to measure the value. Let me reiterate: error is inherent, and is NOT the fault of the person making the measurement. Error is not the same as mistake! Understanding error can greatly improve our effectiveness in making measurements.

Error can be expressed in either "absolute" terms, or using a relative scale. An absolute error would be expressed in terms of "X ± x cm," or some other such unit, while a relative error expression would be "X ± 1% cm." In an electrical circuit, a voltage might be stated as "4.5 volts ± 1%." Which expression to use may be a matter of custom, convention, personal choice or best utility, depending on the situation.

## Categories of Error

There are four general categories of error: theoretical error, static error, dynamic error, and instrument insertion error.

## Theoretical Error

All measurements are based on some measurement theory that predicts how a value will behave when a certain measurement procedure is applied. The measurement theory is usually based on some theoretical model of the phenomenon being measured, i.e. an intellectual construct that tells us something of how that phenomenon works. It is often the case that the theoretical model is valid only over a specified range of the phenomenon. For example, nonlinear phenomenon that have a quadratic, cubic or exponential function can be treated as a straight line linear function over small, carefully selected sections of the range. Electronic sensor outputs often fall into this class.

Alternatively, the actual phenomenon may be terribly complex, or even chaotic, under the right conditions, so the model is therefore simplified for many practical measurements. An equation that is used as the basis for a measurement theory may be only a first order approximation of the actual situation. For example, consider the mean arterial pressure (MAP) that is often measured in clinical medicine and medical sciences research situations. The MAP approximation equation used by clinicians is:

$$\bar{P} = Diastolic + \frac{Systolic - Diastolic}{3} \qquad eq.\ (3\text{-}1)$$

This equation is really only an approximation (and holds true mostly for well people, not some sick people on whom it is applied) of the equation that expresses the mathematical integral of the blood pressure over a cardiac cycle. That is, the time average of the arterial pressure. The actual expression is written in the notation of calculus, which is beyond the math abilities of many of the people who use the clinical version above:

$$\bar{P} = \frac{1}{T} \int_{t1}^{t} P(t)\ dt \qquad eq.\ (3\text{-}2)$$

The approximation (Equation 3-1) works well, but is subject to greater error than Equation 3-2 due to the theoretical simplification of the first equation.

## Static Error

Static errors include a number of different subclasses that are all related in that they are always present even in unchanging systems (thus are not dynamic errors). These errors are not functions of the time or frequency variation.

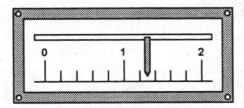

*Figure 3-5. Reading interpolation error.*

Reading Static Error. These errors result from misreading the display output of the sensor system. An analog meter uses a pointer to indicate the measured value. If the pointer is read at an angle other than straight on, then a parallax reading error occurs. Another reading error is the interpolation error, i.e. an error made in judging or estimating the correct value between two calibrated marks on the meter scale (Figure 3-5). Still another reading error occurs if the pointer on a meter scale is too broad, and covers several marks at once.

A related error seen in digital readouts is the last digit bobble error. On digital displays, it is often the case that the least significant digit on the display will flip back and forth between two values. For example, a digital voltmeter might read "12.24" and "12.25" alternately, depending on when you looked at it, despite the fact that absolutely no change occurred in the voltage being measured. This phenomenon occurs when the actual voltage is midway between the two indicated voltages. Error, noise and uncertainty in the system will make a voltage close to 12.245 volts bobble back and forth between the two permissible output states (12.24 and 12.25) on the meter. An example where "bobble" is of significant concern is the case where some action is taken when a value changes above or below a certain quantity - and the digital display bobbles above and below the critical threshold.

Environmental Static Error. All sensors and instruments operate in an environment, which sometimes affect the output states. Factors such as temperature (perhaps the most common error producing agent), pressure, electromagnetic fields, and radiation must be considered in some electronic sensor systems.

Characteristic Static Error. These static errors are still left after reading errors and environmental errors are accounted. When the environment is well within the allowable limits and is unchanging, when there is no reading error, there will be a residual error remaining that is a function of the measurement instrument or process itself. Errors found under this category include zero offset error, gain error, processing error, linearity error, hysteresis error, repeatability error, resolution error and so forth.

Also included in the characteristic error is any design or manufacturing deficiencies that lead to error. Not all of the "ticks" on the ruler are truly 1.0000 mm apart at all points along the ruler. While it is hoped that the errors are random, so that the overall error is small, there is always the possibility of a distinct bias or error trend in any measurement device.

For digital systems one must add to the resolution error a quantization error that emerges from the fact that the output data can only take on certain discrete values. For example, an 8-bit analog-to-digital converter allows 256 different states, so a 0 to 10 volt range is broken into 256 discrete values in 39.06 μV steps. A potential that is between two of these steps is assigned to one or the other according to the rounding protocol used in the measurement process. An example is the weight sensor that outputs 8.540 volts, on a 10-volt scale, to represent a certain weight. The actual 8-bit digitized value may represent 8.502, 8.541, or 8.580 volts because of the ±0.039 volt quantization error.

## Dynamic Error

Dynamic errors arise when the measurand is changing or in motion during the measurement process. Examples of dynamic errors include the inertia of mechanical indicating devices (such as analog meters) when measuring rapidly changing parameters. There are a number of limitations in electronic instrumentation that fall into this category, especially cases where a frequency, phase or slew rate limitation is present.

# Instrument Insertion Error

A fundamental rule of making measurements is that the measurement process should not significantly alter the phenomenon being measured. Otherwise, the measurand is actually the altered situation, not the original situation that is of true interest. Examples of this error are found in many places. One such is the fact that pressure sensors tend to add volume to the system being measured, so slightly reduce the pressure indicated below the actual pressure. Similarly, a flow meter might add length, a different pipe diameter, or turbulence to a system being measured. A voltmeter with a low impedance of its own could alter resistance ratios in an electrical circuit and produce a false reading. This problem is especially seen when using cheap analog volt-ohm-millimeters, that have a low sensitivity hence a low impedance, to measure a voltage in a circuit. The meter resistance $R_m$ is effectively shunted across the circuit resistance across which the voltage appears.

Instrument insertion errors can usually be minimized by good instrument design and good practices. No measurement device has zero effect on the system being measured, but one can reduce the error to a very small value by appropriate selection of methods and devices.

# Dealing with Measurement Error

Measurement error can be minimized through several methods, some of which are lumped together under the rubric "procedure," and others under the legend "statistics."

Under "procedure" one can find methods that will reduce, or even minimize, error contributions to the final result. For example, in an electrical circuit, use a voltmeter that has an extremely high input impedance compared with circuit resistances. The idea is to use an instrument (whether a voltmeter, a pressure meter or whatever) that least disturbs the thing being measured.

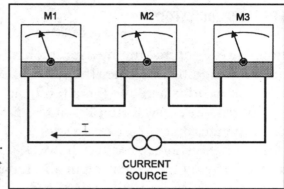

*Figure 3-6. Method for improving measurement accuracy.*

A significant source of measurement error in some electronic circuits is ground loop voltage drops and ground plane noise. This voltage may add or subtract from the reading of output voltage $V_o$ depending on its phase and polarity.

A way to reduce total error is to use several different instruments to measure the same parameter. In Figure 3-6 we see an example where the current flow in a circuit is being measured by three different ammeters: M1, M2 and M3. Each of these instruments will produce a result that contains an error term decorrelated from the error of the others and not biased (unless, by selecting three identical model meters we inherit the characteristic error of that type of instrument). We can estimate the correct value of the current flow rate by taking the average of the three:

$$M_o = \frac{M_1 + M_2 + M_3}{3} \qquad \qquad eq.\ (3\text{-}3)$$

One must be careful to either randomize the system in cases where the sensor or instruments used tend to have large error terms biased in one direction, or calibrate the average error so that it may be subtracted out of the final result.

## Error Contributions Analysis

An error analysis can be performed in order to identify and quantify all contributing sources of error in the system. A determination is then made regarding the randomness of those errors, and a worst case analysis is made.

Under the worst case, one assumes that all of the component errors are biased in a single direction and are maximized. We then attempt to determine the consequences (to our purpose for making the measurement) if these errors line up in that manner, even if such an alignment is improbable. The worst case analysis should be done on both the positive and negative side of the nominal value. An error budget is then created to allocate an allowable error to each individual component of the measurement system in order to ensure that the overall error is not too high for the intended use of the system.

If errors are independent of each other, and are random rather than biased in one direction, and if they are of the same order of magnitude, then one can find the root of the sum of the squares (rss) value of the errors, and use it as a composite error term in planning a measurement system. The rss error is:

$$E_{rss} = \sqrt{\Sigma \, \varepsilon_i^2} \qquad \qquad eq. \ (3\text{-}4)$$

The rss error term is a reasonable estimate or approximation of the combined effects of the individual error components.

A collection of repetitive measurements of a phenomenon can be considered a sampled population, and treated as such. If we take N measurements (M1 through $M_n$) of the same parameter, and then average them we get:

$$\overline{M} = \frac{M1 + M2 + M3 + \ldots + M_n}{N} \qquad \qquad eq. \ (3\text{-}5)$$

The average value obtained in Equation 3-5 is the mean arithmetic average. This value is usually reported as the correct value for the measurement, but when taken alone does not address the issue of error. For this purpose we add a correction factor by quoting the standard error of the mean, or

$$\sigma_{\overline{m}} = \frac{\sigma_m}{\sqrt{N}} \qquad \qquad eq. \ (3\text{-}6)$$

Which is reported in the result as:

$$M = \overline{M} \pm \sigma_{\overline{m}} \qquad \qquad eq. \ (3\text{-}7)$$

Any measurement contains error, and this procedure allows us to estimate that error, and thereby understand the limitations of that particular measurement.

## Operational Definitions in Measurement

Some measurement procedures suggest themselves immediately from the nature of the phenomenon being measured. In other cases, however, there is a degree of ambiguity in the process, and it must be overcome. Sometimes the ambiguity results from the fact that there are many different ways to define the phenomenon, or perhaps no single way is well established. In cases such as these, one might wish to resort to an operational definition, i.e. a procedure that will produce consistent results from measurement to measurement, or when measurements are taken by different people.

An operational definition, therefore, is a defined, standardized procedure that must be followed, and specifies as many factors as are needed to control the measurement so that changes can be properly attributed only to the unknown variable. The need for operational definitions (as opposed to absolute definitions) arises from the fact that things are only rarely so neat, clean and crisp as to suggest their own natural definition. By its very nature, the operational definition does not ask "true" or "false" questions, but rather it asks "what happens under given sets of assumptions or conditions." What an operational definition can do for you, however, is to standardize a measurement in a clear and precise way so that it remains consistent across numerous trials. Operational definitions are used extensively in science and technology. When widely accepted, or promulgated by a recognized authority, they are called standards.

When an operational definition becomes widely accepted, and is used throughout an industry, it may become part of a formal standard or test procedure. You may, for example, see a procedure listed as "performed in accordance with [1]NIST XXXX.XXX" or "ANSI Standard XXX, or AAMI Standard XYZ." These notations mean that whoever made the measurement followed one or another of a published standard.

# Chapter 4

# Instrument Design Rules

The process of designing electronic instrumentation circuits or projects is not an arcane art, open only to a few highly skilled initiates. But rather, it is a logical step-by-step process that can be learned. Like any skill, design skill is improved with practice, so one is cautioned against both excessive expectations "first time at bat" and discouragement if the process did not exactly work out as planned the first time.

Some of the material in this chapter may be called "philosophy," and that may be a fair label to attach. Although many technical people claim to disdain "philosophy," we all have it and use it. It's just that some people think about it a lot, others think about it either a little or at only a few times, while others don't think about it at all - they take actions based not on a logically considered viewpoint, but rather by dumb luck and pure default.

## Design Procedure

The procedure that one adopts to designing may well be different from what is presented below, and that's all right. The purpose of offering a procedure is to systematize the process. While it is conceivable that one can design an instrument by a process similar to Brownian motion, it is the systematic approach that most often yields success. This procedure assumes a product that is a one-of-a-kind instrument, as in a scientific laboratory or plant. Designing a product for production and sale follows a similar procedure, but involves marketing and production problems as well. The steps in the procedure, some of which are iterative with respect to each other, are offered below:

1.  Define and tentatively solve the problem.

2.  Determine the critical attributes required of the final product; incorporate these into a specification and a test plan that determines objective criteria for acceptance or rejection.

3.  Determine the critical parameters and requirements.

4.  Attempt a "block diagram" solution.

5.  Apportion requirements (e.g. gain, frequency response, etc.) to the various blocks.

6.  Perform analysis and do simulations on the block diagram to test validity of the approach.

7.  Design specific circuits to fill in the blocks.

8.  Build and test the circuits.

9.  Combine the circuits with each other on a breadboard.

10.  Test the breadboarded circuit according to a fixed test plan.

11.  Build brassboard that incorporates all changes made in the previous steps.

12.  Test brassboard and correct problems.

13.  Design and construct final configuration.

14.  Test final configuration.

15.  Ship the product.

## Solve the (Right) Problem

The purpose of the designer is to solve some problem or another using analog circuits, digital circuits, a computer or whatever else is available in the armamentarium. There are two related problems often seen in the efforts of novices.

First, it is often the case that the designer will have a tentative favorite approach in mind before the problem is properly understood. Decisions are made based on what the designer is most comfortable doing. For example, many younger designers are likely to select the digital solution in a knee-jerk manner that excludes any consideration of the analog solution. Both should be evaluated, and the one that best fits the need selected.

Second, be sure you understand the problem being solved. While this advice seems trivial, it is also true that failure to understand the problem at hand often sinks designs before they have a chance to manifest themselves. There are several facets to this problem. For example, a natural tendency of engineers is to think that an elegant solution is complex and large-scale. If this mistake is made, then it is likely that the product will be over-designed and have too many whistles and bells. It was, after all, designed to solve a much harder problem than was actually presented.

Another aspect to understanding the problem is to understand the final customer's use for the product. It is all too easy to get caught up in the specification, or our own ideas about the job, and overlook altogether what the user needs to accomplish with the product. An example is derived from biomedical instrumentation. A physiologist requested a pressure amplifier that would measure blood pressures over a range of 0 to 300 mmHg (Torr). What the salesman never told the plant was a) it would be used on humans (safety and regulatory issues), b) that blood would come in contact with the diaphragm (cleaning and/or liquid isolation issues), and c) that it would occasionally be used for measuring 1 to 5 mmHg central venous pressures (which implies low-end linearity issues).

Part of the problem in determining the level of complexity, or the specific design's function, is miscommunication between the end user and the designer. Although miscommunications occur frequently between in-house designers and their clients, it is probably most common between distant customers and engineers in the plant. Of course, marketing people may never let the engineer and customer get together (either from ignorance or a fear that the salesman's little lies will surface: "The reason I hate engineers is that, under duress, they tend to blurt out the truth").

The proper role of the designer is to scope out the problem at hand, understand what the circuit or instrument is supposed to do, how and where the user is going to use it, and exactly what the user wants and expects from the product.

## Determine Critical Attributes

This step is basically the fruit of understanding and solving the correct problem. From the solution of the problem one can determine, and write down, a set of attributes, characteristics, parameters and other indices of the product's final nature.

It is at this point that one must write a specification that documents what the final product is supposed to do. The specification must be clearly written so that others can understand it. A concept or idea does not really exist, except in the mind of the originator. One must, according to W. Edwards Deming, create operational definitions for the attributes of the product. One cannot simply say "it must measure pressure to a linearity of 1-percent over a range of 0 to 100 p.s.i." Rather, it might be necessary instead to specify a test method under which this requirement can be met. There might be, after all, more than one standard for pressure and measurement, and there is certainly more than one definition of linearity. The operational definition serves the powerful function of providing everyone with the same set of rules. Basically, it levels the playing field.

Part of this step, and of making an operational definition, is to write a test plan for the final product. It is here that one determines (and often contractually agrees) exactly what the final product will do, and defines the objective criteria of goodness or badness that will be used to judge the product.

## Determine Critical Parameters and Requirements

Once the product is properly scoped out, it is time to determine the critical technical parameters that are needed to meet the test requirements (and hopefully the user's needs - if the test requirements are properly written). Parameters such as frequency response, gain and so forth tend to vary in multimode instruments, so one must determine the worst case for each specification item and design for it.

## Attempt a Block Diagram Solution

The block diagram is a signals flow or function diagram that represents stages, or collections of stages, in the final instrument. In large instruments there might be several indexed levels of block diagram, each one going finer in detail.

## Apportion Requirements to the Blocks

Once the block diagram solution is on paper, tentatively apportion system requirements to each block. Distribute gain, frequency response, and other attributes to each block. Keep in mind that factors such as gain distribution can have a profound effect on dynamic range. Also, the noise factor and drift of any one amplifier can have a tremendous effect on the final performance - and that it's in these types of parameters (where critical placement of one high quality stage may be sufficient) that added cost and complexity often arises.

## Analyze and Simulate

Once the block diagram is set, and the requirements apportioned to the various stages, it is time to analyze the circuit and run simulations to see if it will actually work. A little "desk checking" goes a long way towards eliminating problems later on when the thing is first prototyped. Plug in typical input values, and see what happens on a stage by stage basis. Check for the reasonableness of outputs at each stage. For example, if the input signal should drive an output signal to, say, 17 volts, and the operational amplifiers are only operated from 5 volt power supplies, then something is wrong and will have to be corrected.

## Design Specific Circuits for Each Block

It is at this point that the remainder of this book is of most usefulness to you, for it is on the specific circuits that we will dwell later. In this step, fill in those blocks with real circuit diagrams.

## Build and Test the Circuits

At this point one must actually construct the individual circuits, and test them to make sure they work as advertised (unless, of course, the circuit is so familiar that no testing is needed). Keep in mind that some of your best ideas for simplified circuits may not actually work - and this is the place to find out. Use a benchtop breadboard that allows circuit construction using plug-in stripped end wires.

## Combine the Circuits in a Formal Breadboard

Once the validity of the individual circuits is determined, combine them together in a formal breadboard. Whether built on a benchtop breadboard (as above), or on a prototyping board, make sure that the layout is similar to that expected in the final form.

## Test the Breadboard

Test the overall circuit according to a formally established objective criteria. This test plan should be developed earlier in step two. As problems arise and are solved, make changes and/or corrections - and document the results. It is, perhaps, the main failing of inexperienced designers that they do not properly document their work, even in an engineer's or scientist's notebook.

# Build and Test a Brassboard Version

A "brassboard" is a version made as close as possible to the final configuration. While breadboarding techniques can be a little sloppy, the brassboard should be a proper printed circuit board. The test criteria should be the same as before, updated only for changes that occurred. If problems turn up, they should be corrected prior to proceeding further. Keep in mind that the most common problems that occur in leaping from breadboard to brassboard are layout (e.g. coupling between stages), power distribution and ground noise (these are the areas of difference between the two configurations).

# Design, Build and Test Final Version

Once all of the problems are known, solved and changes incorporated, it is time to build the final product as it will be given to the end user. It is at this point that they reputation of the designer is made or broken, because it is here that the product is finally evaluated by the client.

# Chapter 5

# Resistive Sensors

Sensors are electronic or electrical devices that change in some significant way in response to an applied stimulus. One common class of sensor is the resistive sensor. We find in this class a variety of position sensors, angle sensors, pressure sensors, temperature sensors and light sensors, all based on slightly different physical phenomena. We will deal with temperature sensors and light sensors in subsequent chapters.

## Potentiometer Sensors

Resistors come in a variety of forms, one of which is the potentiometer. The "pot" (Figure 5-1) is a variable resistor with three terminals and an actuator shaft that can change the position of a tap on the resistor body. The total resistance (R) is measured between the two outside terminals, while the fraction of R from one outside terminal to the center terminal is $R_A$ and from the other to the center terminal is $R_B$. The center terminal is connected to a tap that rides on the resistive element. In terms of Figure 5-1, the overall resistance is R. The resistance from the left terminal to the center terminal is $R_A$, which is equal to R - $R_B$. Similarly, $R_B$ = R - $R_A$.

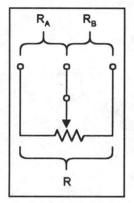

*Figure 5-1.*
*Potentiometer sensor.*

The term "potentiometer" comes from the late 19th century when it was used as a precision means of providing reference voltages. A highly accurate voltage source, such as a Weston cell, was connected across the outside terminals. By knowing the position of the tap, we also know the value of the voltage appearing between one end of the pot and the wiper.

Several different forms of element are available for use in potentiometers: wire, carbon composition, and metal film. The wire form uses a resistive wire wrapped uniformly around a support. The carbon composition type uses a flat surface on which the appropriate carbon composition material is deposited. The metal film type is conceptually similar to the carbon form, but a metallic film is deposited on a ceramic substrate.

The taper of the potentiometer refers to the profile of resistance change with change of tap position. The most commonly used for sensor applications is the linear taper potentiometer. These devices have the same change of resistance ($\Delta R$) for any unit change of tap position ($\Delta X$). The audio taper is shaped to optimize the use of the potentiometer as a audio volume control. The idea is to get a smooth transition of volume (the human ear is markedly nonlinear). Still another form is the logarithmic taper, i.e. the change of resistance changes as the logarithm of the position change.

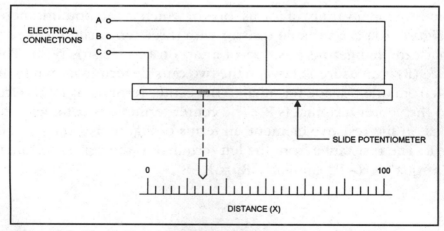

*Figure 5-2. Linear ("straight line" or "slide") potentiometer.*

Figure 5-2 shows one form of potentiometer often seen in sensor applications: the slide or linear potentiometer. The body of the potentiometer is a rectangular shape, and has a slot cut into one side for nearly the entire length. The slot accommodates the actuator for the wiper. If we couple the wiper actuator to some external mechanical device that changes in position "X" (represented here by a pointer and a scale), then the position of the wiper actuator gives an indication of the position of the external device. In some cases, mechanical linkage is used to reduce the mechanical translation distance to the "throw" of the actuator wiper.

Figure 5-3 shows the equivalent in a rotary shaft potentiometer. These devices are the most familiar form of pot for most readers, I suspect. The resistive element is arranged in a circular pattern coaxial to the actuator shaft. The respective values of R, $R_A$ and $R_B$ provide the angular position of the shaft.

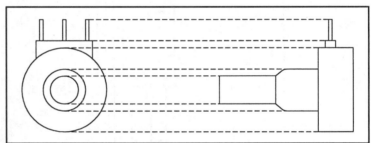

*Figure 5-3. Normal rotary potentiometer.*

Several different forms of angular potentiometer are available. Most common forms have a resistive element that covers 270° of the circuit (although I've seen 310° pots as well). Some of them don't have stops at either end of the range, but have 360° rotation. Some of these pots have a fourth terminal at the crossover point to indicate zero position. Another variety is the multi-turn potentiometer. These devices cover the entire resistance range in five, ten, fifteen or twenty turns of the shaft.

A number of years ago I worked repairing medical electronic devices. One of those devices was a syringe pump used in a research laboratory. The pump had a worm gear mechanism and a holder for a 50 cc syringe (one of those big honkin' jobs you hate to see the doctor coming at you with!). A precision step motor turned the worm gear, moving a push plate against the back of the syringe (shown simplified in Figure 5-4). A linear potentiometer was ganged to the worm gear (or in version to the push plate).

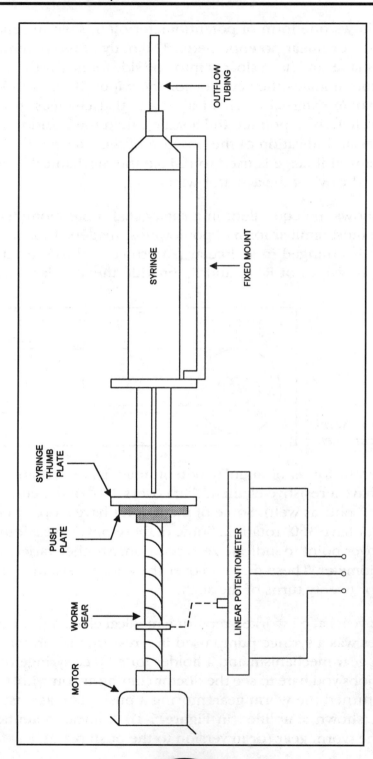

*Figure 5-4. Slide potentiometer used to measure the flow from a syringe in a syringe pump.*

The syringe pump potentiometer is a displacement sensor, i.e. its resistance provides information on the position of the wiper, hence in turn the position of the plunger of the sensor. By simple ratio and proportion (assuming linear taper potentiometer is used), we can figure out the volume of material inside the syringe has been delivered ($R_A$), i.e. flow volume, and how much is left ($R_B$) in the syringe. Indirectly, we can also measure

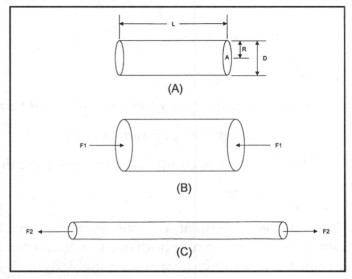

*Figure 5-5. Wire element a) at rest; b) with compression force applied; c) with tension force applied.*

flow rate because flow rate is flow per unit of time. We can take a smoothed measure of flow rate by performing a little calculus on flow volume.

As an aside, we can often indirectly measure various parameters by taking either the derivative or integral of some related parameter. For example, a potentiometer displacement sensor provides an indication of position X. If we take the first derivative of X (i.e. $\Delta X / \Delta T$) we get the velocity, and if we take the second derivative of X (or first derivative of velocity) we get acceleration. Similarly, integrating flow rate gives us flow volume. This is one of the uses of integrator and differentiator circuits.

## Piezoresistive Strain Gages

The resistance of any specific conductor is directly proportional to its length and inversely proportional to its cross-sectional area (see Figure 5-5). Resistance is also directly proportional to a property of the conductor material called resistivity. Equation 5-1 shows clearly that resistance is proportional to resistivity and the length of the conductor, and inversely proportional to cross-sectional area of the conductor.

● $\quad R = \dfrac{\rho L}{A}$ *eq. (5-1)*

Where:

ρ is the resistivity in ohm-centimeters (Ω-cm).

L is the length in centimeters (cm).

A is the cross-sectional area in square centimeters

(cm²), or $\pi R^2$.

The word piezoresistivity denotes the resistance change that takes place when either the length, area or both of a conductor are changed. Figure 5-5a shows a cylindrical conductor with an initial length $(L_o)$ and a cross-sectional area $(A_o)$.

When a compression force (F1) is applied, as in Figure 5-5b, the length reduces and the cross-sectional area increases. This situation results in a decrease in the electrical resistance. Mathematically:

● $\quad R = (R_o - \Delta R) \propto \left(\dfrac{L_o - \Delta L}{A_o + \Delta A}\right)$ *eq. (5-2)*

Similarly, when a tension force is applied (Figure 5-5c) the length increases and the cross-sectional area decreases, so the electrical resistance will increase.

● $\quad R = (R_o + \Delta R) \propto \left(\dfrac{L_o + \Delta L}{A_o - \Delta A}\right)$ *eq. (5-3)*

In either tension or compression cases, provided that the physical change is small, the change of electrical resistance is a nearly linear function of the applied force, so can be used to make measurements of that force. Sensors that use piezoresistivity to measure forces are called strain gages.

## Strain Gages

A strain gage is a piezoresistive element, either wire, metal foil or semiconductor, designed to create an electrical resistance change when a force is applied. Strain gages can be classified as either bonded or unbonded types. Figure 5-6 shows both methods of construction.

The unbonded strain gage is shown in Figure 5-6a, and consists of a wire resistance element stretched taut between two flexible supports. These supports are configured in such a way as to place a tension or compression force on the taut wire when external forces (F) are applied. In the particular example shown, the supports are mounted on a thin metal diaphragm that flexes when a force is applied. A tension force will cause the flexible supports to spread apart, placing increased tension force on the wire and thereby increasing its resistance. Alternatively, when a compressing force is applied, the ends of the supports tend to move closer together, effectively placing a compression force on the wire element and thereby reducing its resistance. In actuality, the wire's resting condition is tautness, which implies a tension force, so "tension" and "compression" mean an increase or decrease in normal tension, respectively.

The bonded form of strain gage is shown in Figure 5-6b. In this type of device a wire, foil or semiconductor element is cemented to a thin metal diaphragm. When the diaphragm is flexed, the element deforms to produce a resistance change.

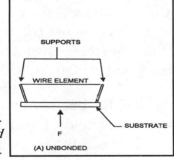

*Figure 5-6a. Unbonded strain gage.*

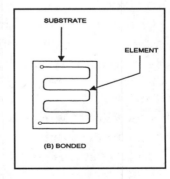

*Figure 5-6b. Bonded strain gage.*

The linearity of both types of strain gage can be quite good, provided that the elastic limits of the diaphragm and element are not exceeded. It is also necessary to ensure that the change of length is only a small percentage of the resting length.

In the past, the "standard wisdom" held that bonded strain gages are more rugged, but less linear than unbonded models. Although this may have been the situation at one time, recent experience has shown that modern manufacturing techniques can produce rugged, linear, reliable units of both types of construction.

## Strain Gage Sensitivity

The sensitivity of the strain gage is expressed in terms of unit change of electrical resistance per unit change of length, and is most often given in the form of the gage factor (S) for the element:

$$S = \left(\frac{\Delta R / R}{\Delta L / L}\right) \qquad \text{eq. (5-4)}$$

## Resistive Sensor Circuitry

The resistance of the resistive sensor can be used for making the measurement in question, but that is not usually the best approach. A better approach is to use the resistive sensor to generate a voltage that is proportional to the applied stimulus. One way to do that trick is to connect the sensor in a voltage divider network. These circuits consist of a fixed resistor (R1) and the sensor resistance (R2) in series across an excitation voltage (V). The output voltage ($V_o$) is taken across the sensor, so is equal to $(V \times R2)/(R1 + R2)$. The problem with this approach is that the voltage will always have some non-zero value when the applied stimulus is zero, unless the resistance of the sensor goes to zero (or very near it) at the same time. Lots of luck finding that situation. A better way is to use a Wheatstone bridge circuit.

Figure 5-7 shows the classic Wheatstone bridge. It's been around since the mid-19th century, but still forms the basis of a gozillion electronic instruments. One way to look at this circuit is as a pair of voltage dividers: R1/R2 forms one and R3/R4 forms the other. If a voltage is applied across these

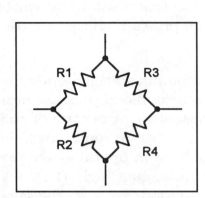

Figure 5-7. Wheatstone bridge.

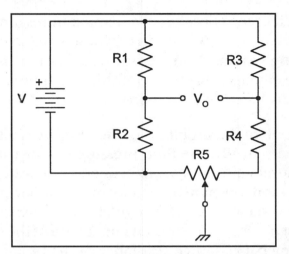

Figure 5-8. Redrawn Wheatstone bridge with balancing potentiometer.

voltage dividers (e.g. one side of the battery connected to junction R1-R3, and the other to junction R2-R4), then the output voltage appears across the opposite nodes (e.g. junctions R1-R2 and R3-R4).

Consider the situation where junction R2-R4 is grounded. In this case, the output voltage is the difference between the voltage drop across R2 and the voltage drop across R4: $V_o = V_{R2} - V_{R4}$. The nice thing about this circuit is that the output voltage can be made zero under zero stimulus by selecting resistor values. The output voltage is zero when the ratios of the two voltage dividers are equal to each other: R1/R2 = R3/R4.

A more practical version of the Wheatstone bridge is shown in Figure 5-8. In this circuit a balancing potentiometer (R5) is connected between R2 and R4 at junction R2-R4. The value of R5 is usually a fraction of R2 or R5, which are set equal to each other in most cases. By adjusting R5 we can zero-out any offsets due to minor variations in the resistances of R1-R4. The adjustment process is to set the applied stimulus to zero, and then adjust R5 for $V_o = 0$.

Different forms of sensor use different variations on the bridge theme. In some circuits, only one resistor (usually R2 or R4) is a sensor. Some temperature sensors or displacement sensors work this way. In other cases, two resistors (e.g. R2 and R4) will be sensors, and the other two fixed. One example of this is the differential thermometer, i.e. a thermometer that measures the difference between two temperatures. An indoor-outdoor thermistor pair is used in this way on an environmental temperature controller for home heating systems. Many sensors are available in which all four resistors are sensor elements. This is the most linear approach. Fluid pressure sensors are usually based on this approach.

One of the problems of the bridge circuit is that output voltages tend to be very small. One fluid pressure sensor used in human blood pressure measurements, for example, offers an output potential of 50 mV per volt of excitation per mmHg of pressure. Another problem is that, in some cases, the resistances of the bridge elements are quite high (as in optical or thermal sensors), so the output impedance of the circuit is also quite high. The "looking back" source resistance of the bridge, when all four arms are the same resistance, is the resistance of any one element (if you doubt this, apply Thevinin's theorem). Clearly an amplifier is in order for most sensors.

Figure 5-9 shows the standard amplifier circuit for Wheatstone bridge sensors. It is a differential amplifier made with one operational amplifier. When R4 = R5 and R6 = R7, the gain of the amplifier is R7/R4. The amount of gain to use depends on the application. The rule-of-thumb is that the input resistors R4 and R5 should be at least ten times the looking back resistance of the bridge. A number of modern bridge sensors take advantage of modern integrated circuit technology to build the amplifier right into the housing of the sensor.

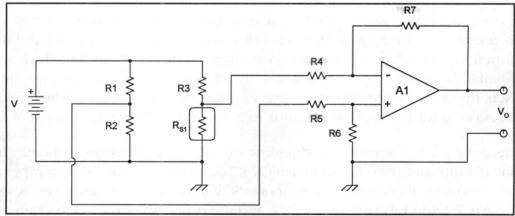

*Figure 5-9. Differential amplifier used to boost output voltage from Wheatstone bridge, while decreasing source resistance.*

# Chapter 6

# Temperature Sensors

In this chapter we will look at the different forms of temperature sensors and the thermometry (e.g. measurement of temperature) circuits that support them. The idea is not so much to present a comprehensive treatment, but rather an overview of the topic.

## Temperature Scales

There are several different scales used in the measurement of temperature. The familiar Fahrenheit and Celsius (a.k.a. Centigrade), along with the less familiar Kelvin and Rankine scales are used. The Celsius and Fahrenheit scales are arranged such that 0 °C is the same temperature as 32 °F. Both are defined by the "ice-point" (i.e. the freezing point of water at standard temperature and pressure). The two scales can be converted to each other by the equation:

$$°F - 32 = 1.8°C \qquad\qquad \text{eq. (6-1)}$$

Where:

F is degrees Fahrenheit (°F).

C is degrees Celsius (°C).

The Kelvin scale uses the same size degree steps as the Celsius scale, but defines the zero degree point differently. In the Celsius scale, 0 °C is the ice-point, while in the Kelvin scale it is Absolute Zero (the point where molecular activity ceases). Thus, 0 K » -273.16 °C. The Celsius and Kelvin scales are converted by:

$$°K = °C + 273.16 \qquad \qquad eq. (6\text{-}2)$$

$$°C = °K - 273.16 \qquad \qquad eq. (6\text{-}3)$$

The Rankine scale is to the Fahrenheit scale as the Kelvin scale is to the Celsius. In other words, the size of Rankine degree steps are the same size as Fahrenheit degrees, but the zero point is at absolute zero on the Fahrenheit scale. Thus, 0 °R is approximately equal to -459.7 °F.

Several different sensors are commonly used to measure temperature: Thermal resistors (RTDs and thermistors), thermocouples and PN semiconductor junctions. Although applications for these different forms of sensor overlap, there are often key parameters and other factors that favor one or the other. Let's examine each of these sensor types in turn.

## Thermal Resistors

Thermistors are electrically conductive elements that are designed to change electrical resistance in a predictable manner with changes in applied temperature. There are two basic classes of thermal resistors, resistance temperature devices (RTD) and thermistors.

The amount of resistance change is designated by the temperature coefficient (a) of the material, which is measured in ohms of resistance change per ohm of resistance per degree Celsius. A positive temperature coefficient (PTC) device increases resistance with increases in temperature. Alternatively, negative temperature coefficient (NTC) devices decrease resistance with increases in temperature. Typical curves for NTC and PTC thermistors, and a platinum RTD, are shown in Figure 6-1. The usual circuit symbols for thermal resistors are shown in Figure 6-2. The indirectly heated variety uses an internal heating element, while the directly heated form takes its heating from the environment.

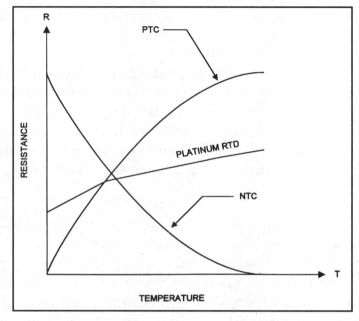

*Figure 6-1. Positive and negative temperature coefficient curves.*

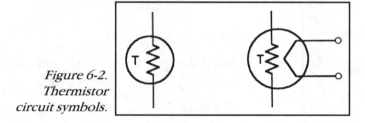

*Figure 6-2. Thermistor circuit symbols.*

## Resistance Change with Temperature

Most thermal resistors have a nonlinear characteristic curve when it is plotted over a wide temperature range. But when limited to narrow temperature ranges, the linearity is considerably better (again see Figure 6-1). When such devices are used, however, it is necessary to ensure that the temperature will not go on excursions outside of the permissible linear range. There are methods for linearizing the devices, and these will be discussed in a later section.

Thermal resistors are among the oldest temperature sensors known. The temperature sensitivity of electrical resistance in silver sulfide was noted by physicist Michael Faraday in 1833, and that of iron wire was equally well known at early dates. There are several different types of thermal resistor, but the simplest is the resistance temperature device (RTD). The RTD may be either a wire element or a thin metallic film.

Simple RTD elements are based on the tendency of materials to change physical dimensions with changes in temperature. Metals, for example, tend to expand when heated. For metal wires, the resistance is directly proportional to the length of the sample. Thus, when a metal is heated, it tends to expand so its electrical resistance increases. Most metals have a positive temperature coefficient (a > 0). Copper, for example, has a temperature coefficient of +0.004.

Not all materials have positive temperature coefficients, however. Some materials, like carbon and some ceramics, have a negative temperature coefficient (e.g., for carbon a = -0.0005). Other materials, including certain metal alloys or oxides, have temperature coefficients that are very low. For example, in manganin and constantan the temperature coefficient is approximately +0.00002, and for nichrome it is +0.00017. Table 6-1 shows the temperature coefficients of some common materials.

Table 6-1

| Material | Resistivity ($\rho$) $\Omega \bullet cm$ | Temperature ($\alpha$) Coefficient ($\Omega/\Omega/°C$) |
|---|---|---|
| Carbon (C) | 3,496 | -0.0005 |
| Iron (Fe) | 10 | +0.005 |
| Nichrome | 101 | +0.00017 |
| Platinum | 10 | +0.00377 |
| Silver | 1.628 | +0.0038 |
| Aluminum (Al) | 2.828 | +0.0036 |
| Annealed Copper | 1.724 | +0.0039 |
| Gold | 2.4 | +0.004 |
| Nickel | 6.84 | +0.0067 |
| N-Type Silicon | 1.43 | +0.007 |
| P-Type Silicon | 0.62 | +0.007 |

The change in resistance caused by changes in temperature is a function of a and the value of temperature change. For a wire element, the new resistance is found from:

$$R_{T2} = R_{T1}[1 + \alpha(T2 - T1)] \qquad eq.\ (6\text{-}4)$$

Where:

$R_{T1}$ is the starting temperature resistance.

$R_{T2}$ is the final temperature resistance.

$\alpha$ is the temperature coefficient.

T1 is the starting temperature.

T2 is the final temperature.

In some sources, $R_{T1}$ is replaced with $R_o$ when temperature T1 is 0 °C. In that case, T1 is relabelled $T_o$.

Wire elements are sometimes used as thermal resistors. Platinum wire RTD elements, for example, are frequently used as sensors at temperatures down to 20 K, and rhodium is used below 20 K. In platinum RTDs the resistance change with temperature ($\Delta R/\Delta T$) is about 0.377%/°C, with linearity of ±0.2% over the range 0 °C to +100 °C, and accuracies of ±0.001 °C.

The platinum RTD is used as an international standard in thermometry. One International Platinum Temperature Scale (IPTS) defines a standard platinum RTD over the range from the boiling point of oxygen ($O_2$), -183.962 °C, to the melting point of antimony, +630.74 °C.

In 1968, a new standard (IPTS-68) was specified in which the reference end points are the triple point of hydrogen (13.81 K) and the freezing point of antimony (+903.89 K). The general expression below applies:

$$R_T = R_o[1 + a_1 T + a_2 T^2 + a_3 T^3 + ... + a_n T^n] $$

$$eq.\ (6\text{-}5)$$

# Electronic Circuit Guidebook, Volume 1: Sensors

For the limited temperature range 0 °C to +100 °C, a simplified equation is often used:

$$R_T = R_o(1 + a_1 T)$$   eq. (6-6)

and over 0 °C to +630 °C:

$$R_T = R_o + \alpha R_o T + \beta R_o T^2$$   eq. (6-7)

Where:

$R_T$ is the thermistor resistance at temperature T.

$R_o$ is the resistance of the thermistor at the ice point (0 °C).

T is the temperature being measured.

$\alpha$ is a constant (3.98 × 10⁻³ Ω/°C for platinum).

$\beta$ is a constant (-0.586 × 10⁻⁶ Ω/°C for platinum).

Another expression that is frequently used in thermal resistor calculations over the old IPTS range (-183 to +630 °C) is the Callendar-Van Dusen equation:

$$R_T = R_o \alpha [T - \sigma(0.01T - 1)(0.01T) - \beta(0.01T - 1)(0.01T)^3]$$

   eq. (6-8)

Where:

$R_T$ is the resistance at temperature T.

$R_o$ is the resistance at the ice-point (0 °C).

$\alpha$ is the resistance at 100 °C.

$\beta$ is the thermistor resistance at -183.96 °C.

$\sigma$ is the resistance at +444.7 °C.

Typically, $\alpha$ = 3.9 X 10-3, $\beta$ = 0 if T > 0 or 0.11 if T < 0, $\sigma$= 1.49 when $R_o$ = 100).

# Thermistors

Another type of thermal resistor (the thermistor) is made of evaporated films, carbon or carbon compositions, or ceramic like semiconductors formed of oxides of copper, cobalt, manganese, magnesium, nickel, titanium or uranium. Unlike the basic RTD device, thermistors can be molded or compressed into a variety of clever shapes to fit a wide range of applications (Figure 6-3). These devices have a resistance change characteristic of 4 to 6 %/°C with generally a negative temperature coefficient. A special class of thermistor, called posistors, which are made of barium titanate or strontium titanate ceramics, have a positive temperature coefficient. Positive temperature coefficients are also found in silicon thermistors in which the Si semiconductor is doped to a density of about $10^{16}/cm^3$.

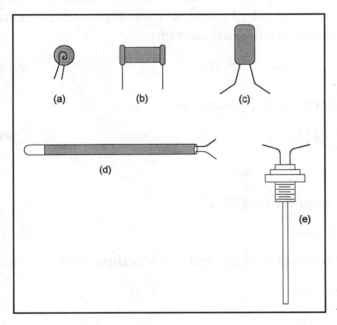

*Figure 6-3. Thermistor package types.*

# Thermistor Parameters

Before one can successfully use thermistors of either type it is necessary to first understand some of the basic properties of the thermistor. These are expressed in the form of certain standard parameters.

Cold (Zero-Power) Resistance. This parameter is the resistance of the thermistor at a standard reference temperature, usually either room temperature (25 °C or the ice-point of water - 0 °C) under conditions of no self-heating power dissipation. This parameter is the cold resistance that is listed in the specifications sheet as the nominal resistance. For example, a device listed as a "1000 ohm thermistor" has a resistance of 1000 ohms at the standard reference temperature (25 °C unless otherwise specified). The conditions under which the thermistor is operated for measurement of the cold resistance include a requirement that the current through the device be sufficiently low to avoid self-heating.

Hot Resistance. The hot resistance of the thermistor is measured when the device is operated at a higher temperature than the cold resistance temperature. The higher temperature is due to ambient temperature, the current flow through the thermistor, the applied heater current (indirectly heated types only), or a combination of all of these factors. Equation 6-4 can be modified to find the hot resistance of wire elements:

$$R_T = R_o[1 + \alpha(T - T_o)] \qquad \text{eq. (6-9)}$$

For other forms of thermistor the expression is:

$$R_T = R_o\, e^{\beta\left(\frac{1}{T} - \frac{1}{T_o}\right)} \qquad \text{eq. (6-10)}$$

Where:

$T_o$ is the reference temperature (25 °C).

T is the new temperature.

$R_o$ is the thermistor resistance at the reference temperature.

$R_T$ is the resistance at temperature T.

$\alpha$ is the coefficient of resistance.

$\beta$ is a materials factor called the characteristic temperature, with units of degrees Kelvin (usually between 1500 K and 7000 K, typically 4000 K).

Resistance vs. Temperature. This parameter is an expression of the character-istic curve shown in Figure 6-1. The exact shape of the curve is a function of the thermistor in question, but it will be of the form shown in Figure 6-1 and is quite nonlinear in certain parts of its range.

Resistance Ratio ($R_r/R_o$). The resistance ratio is essentially a simplified ex-pression of the R vs. T curve. It states the ratio of the thermistor resistance at a specified temperature (50 °C, 100 °C, or 125 °C) to the cold temperature (25 °C) resistance.

Voltage vs. Current (V vs. I). Directly heated thermistors have an unusual voltage vs. current curve that includes both ohmic and negative resistance regions. Assuming a constant ambient temperature, an increase in current through the thermistor will cause a linear increase in voltage drop across the thermistor. Because this behavior is in accordance with Ohm's law, $V = IR$, that portion of the curve is called the ohmic or positive resistance region. At a certain point, however, internal self-heating becomes dominant and begins to alter the resistance of the thermistor. At this point the voltage drop begins to decrease with increasing current flow. In other words, in this region the thermistor is a negative resistance device.

Maximum Power ($P_{max}$). This parameter is the maximum allowable constant power level that the thermistor will handle without either destruction, per-manent alteration of its characteristics, or degradation of its performance.

Dissipation Constant ($\delta$). This factor, symbolized in the specifications sheet by Greek delta ($\delta$), is the ratio of the change in power dissipation for small changes in the body temperature of the thermistor ($\delta = \Delta P_d/\Delta T_B$).

Sensitivity ($\gamma$). The sensitivity of a thermistor is the ratio of resistance change to temperature change ($\Delta R/\Delta T$), expressed as a percent change per degree of temperature. Because the R vs. T curve (Figure 6-1) is nonlinear over most of its range, sensitivity factor numbers are valid only over a limited range. Typi-cal values run from 0.5%/°C to 4%/°C.

Temperature Range (including $T_{min}$ and $T_{max}$). The thermistor's characteris-tics are only specified over a limited temperature range, $T_{min}$ to $T_{max}$. The value of $T_{min}$ is typically -200 °C, while $T_{max}$ is typically +650 °C (although there are devices with a narrower range).

Thermal Time Constant (t). The body temperature of a thermistor does not change instantaneously in response to a step-function change in ambient temperature. If $T_i$ is the initial temperature, and $T_f$ is the final temperature, then the thermal time constant is the time required for the body temperature of the thermistor to change 63.2 percent of the range between these two temperatures. The term "63.2" is derived from $[1 - e^{-t}]$ when $t = 1$ second.

## Linearizing Thermistors

The R vs. T curve seen earlier in Figure 6-1 is nonlinear over much of its range. For some measurements, therefore, it is necessary to either restrict the use of the device to a limited range of temperatures, or to actually linearize the R vs. T curve. There are several ways to linearize the curve. Some of them involve electronic circuits, so will be discussed in detail after we have discussed the circuits involved. There are, however, two methods that only involve simple resistors or other thermistors. Figure 6-4a shows several linearization networks used by a thermistor manufacturer. Although the network functions (to an outside observer) like a single, two-terminal thermistor, it actually consists of a network of resistors and thermistors.

A relatively easy method for linearizing a thermistor is shown in Figure 6-4b. This method involves shunting a low temperature coefficient resistor, $R_s$, across the thermistor, $R_t$. The total value of the network is the parallel resistance of the two elements:

$$R_{total} = \frac{R_t R_s}{R_s + R_s} \qquad eq.\ (6\text{-}11)$$

The value of Rs is the mean value ($R_m$) of $R_t$ over the temperature range of interest. Suppose, for example, that you want to linearize a thermistor over the human physiological temperature range (e.g. 30 °C to 45 °C). The value of $R_m$ in this case, hence the value of $R_s$, is the thermistor resistance at a temperature of (30 + (45-30)/2) °C, or 37.5 °C. Figure 6-4c shows the relationship between the unlinearized thermistor resistance ($R_t$) and the parallel fixed resistor $R_m$. Note that $R_{total}$ is considerably more linear over a larger portion of its temperature range than the unlinearized version.

The expression for the value of the total resistance $R_{t'}$ is:

$$R_{t'} = \frac{R_m}{2}\left(1 + \frac{\alpha}{2}(T - T_m)\right)$$

eq. (6-12)

There are other methods for linearizing thermistors, but these are not discussed here.

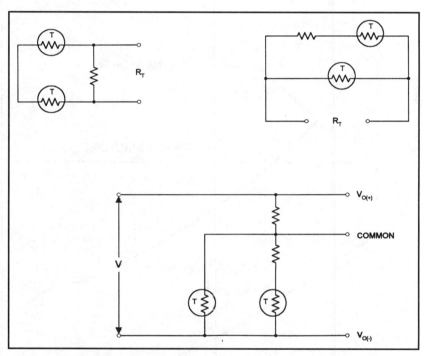

Figure 6-4a. Thermistor linearization circuits.

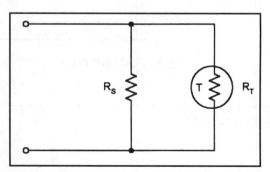

Figure 6-4b. Simplified circuit.

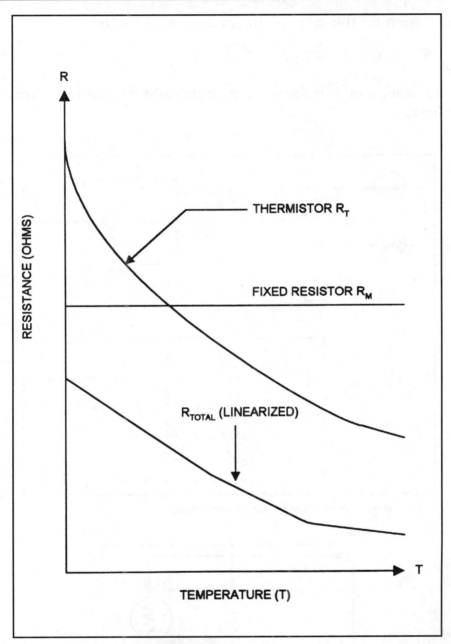

*Figure 6-4c. Regular and linearized curves.*

# Thermal Resistor Error

Thermal resistors are sensitive to certain forms of error. The self-heating error was discussed earlier. It is controlled by making the current in the thermistor below a certain critical value at which self-heating occurs. Self-heating is also controlled by using a higher current, but presenting it in the form of low duty-cycle pulses rather than a dc level. At the output of the thermography circuit one can either synchronously sample the output voltage in step with the pulses, or integrate the pulses to form a dc level that is proportional to the average of the pulse train amplitudes.

Another source of error is thermoelectric potentials in the circuit. Any time two different metallic surfaces are brought into contact, as when wires and connection terminals are joined together, a small electromotive force (emf) is created. Below you will find that this emf is the basis for thermocouple temperature sensors. But in a thermal resistor circuit the potential must be controlled. Perhaps the best way to measure it, where it is a factor, is to measure the potential and then subtract it out of the final result. Some electronic circuits do this automatically.

Another potential error exists because of the resistance of the leads that make up the thermal resistor assembly. Unfortunately, many thermal resistors, especially RTDs, have very low resistance values. These values are close to the typically very low resistances found in the leads that connect the thermal resistor to its circuits.

The lead resistance ($R_L$) error can be reduced by separating the lead carrying the excitation current from the output voltage lead. A three-wire bridge circuit will accomplish this job. Make all three wires heavy with respect to the amount of current that they carry in order to reduce the voltage drops associated with $R_L$.

# Thermocouples

Thermoelectricity is the property of generating an electrical potential from heat. Thermoelectricity was discovered by Thomas J. Seebeck in 1821 during experiments on electromagnetics in circuits containing bismuth and either copper or antimony. The discovery of Seebeck was that a junction made of two different metals will produce an electromotive force (emf) when it is heated (Figure 6-5a). Within a decade, A.C. Becquerel discovered that thermoelectricity could be used to measure temperature. Today we call the Seebeck/Becquerel junction a thermocouple. When several thermocouples are connected together in series the combination is called a thermopile (Figure 6-5b). Several thermocouples can be connected in parallel to find the arithmetic mean of several temperatures (Figure 6-5c). When a thermocouple system is connected to a resistive load, a current will flow that is proportional to the thermoelectric potential, and inversely proportional to the resistance. By convention, the current in the circuit is positive if it flows from hot to cold.

When the junction of dissimilar metals is heated, an emf called the Seebeck potential $(V_s)$ of the system is created. This potential is related to the temperature and to the difference in the work functions (difficulty in stripping electrons from their associated metallic atoms) of the two metals. The Seebeck potential varies from about 6 μV/°C to about 90 μV/°C, and is found by integrating:

$$dV_s = \alpha_{a,b} dT \qquad \qquad eq. (6-13)$$

Where:

   $V_s$ is the Seebeck potential.

   $\alpha_{a,b}$ is the Seebeck coefficient of the system (see below).

   T is the temperature.

The Seebeck coefficient is determined empirically by measuring the coefficient for both materials (A and B) against a third material, called the reference material. Thus, one must find $\alpha_{ar}$ and $\alpha_{br}$ at a standard reference temperature, and then take their algebraic sum. We can further define the Seebeck coefficient in the form:

$$\alpha_{a,b} = \frac{dV_s}{dt} \qquad \qquad eq. (6-14)$$

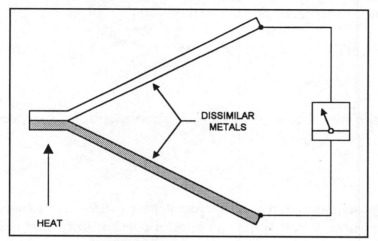

*Figure 6-5a.*
*Thermocouple.*

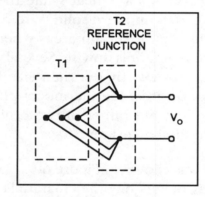

*Figure 6-5b.*
*Thermopile.*

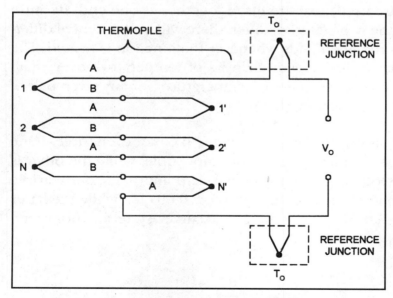

*Figure 6-5c.*
*Advanced thermopile.*

Another expression for the Seebeck potential is:

$$V = \alpha(T1 - T2) + \gamma(T1^2 - T2^2) \qquad \text{eq. (6-15)}$$

Where: $\alpha$ and $\gamma$ are constants.

The sensitivity or thermoelectric power of the thermocouple is expressed by:

$$S = \frac{dV}{dT} = \alpha + 2\gamma T \qquad \text{eq. (6-16)}$$

A subsequent study by Jean Charles Althanase Peltier (1836-35) showed an interesting phenomenon. When Peltier inserted a current into the Seebeck circuit, he found that one junction absorbs heat while the other gives up heat. This effect is the basis for certain modern solid-state refrigerators used in recreational vehicles, camping and in certain areas where the noise and bulk of a compressor operated refrigerator is unwise. Several companies make styrofoam electrical coolers that are exactly the right size for a six-pack of beer or soda. In practical Peltier devices, the thermocouple consists not of two different metals fused together, but rather of two regions of semiconductor differentially doped with dissimilar impurities.

An example of a thermocouple was shown in Figure 6-5a. This type of transducer consists of two dissimilar metals or other materials (some ceramics and semiconductors are used) that are fused together at one end. Because the work functions of the two materials differ, there will be a potential difference generated across the open ends whenever the junction is heated. The potential is approximately linear with changes of temperature over small ranges, although over very large ranges of temperature (for any given pair of materials) nonlinearity increases markedly.

Thermocouples are typically used in pairs (Figure 6-6a) or even threes. One junction will be used as the measurement thermocouple, while the other is the cold junction. The cold junction may be the ice-point (0 °C), or the triple point of $H_2O$. The former gives accuracy of 0.05 ± 0.001 °C, while the latter is capable of 0.01 ± 0.0005 °C. In some low accuracy systems, room temperature may be used.

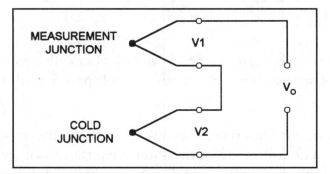

*Figure 6-6a. Two junction thermocouple measures temperature junction against a cold junction.*

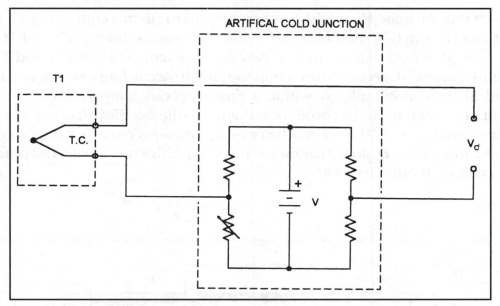

*Figure 6-6b. Artificial cold junction.*

The difference in Seebeck potentials produced by the two junctions is used as a measure of the temperature difference between them. In laboratory settings where the ice-point is used, the cold junction might be an ice water bath. A dewar flask is filled with water and ice chips, and then stoppered. A measurement thermistor and a reference junction are immersed in the bath to produce the cold junction potential. Although it can be successfully argued that the ice water bath is the best cold junction for high accuracy work, the use of an artificial cold junction (Figure 6-6b) is popular. In the artificial cold junction, a potential equal to the expected ice-point potential is applied to the circuit in lieu of the second thermocouple.

## Thermoelectric Laws

Over the nearly two centuries since Thomas Seebeck discovered thermoelectricity a series of several "laws" have been developed regarding the phenomena.

Law of Homogeneous Materials. This law states that a thermoelectric current will not flow in a circuit made of the same material. In other words, unless the materials of the conductors are dissimilar, i.e. have different work functions, then no current is generated by heating the junction.

Law of Intermediate Materials. This law pertains to circuits containing a third material (Figure 6-7). In this circuit, a pair of thermocouples (TC1 and TC2) are formed of materials A and B, at two different temperatures (T1 and T2). A third material (C) is introduced into the circuit. According to this law, if the third material is completely within a homogeneous temperature environment (T3), then it will not produce a change in the net emf. In other words, in terms of Figure 6-7, there will be two A-C junctions entirely within a constant temperature region. That means their respective thermoelectric potentials null each other to zero.

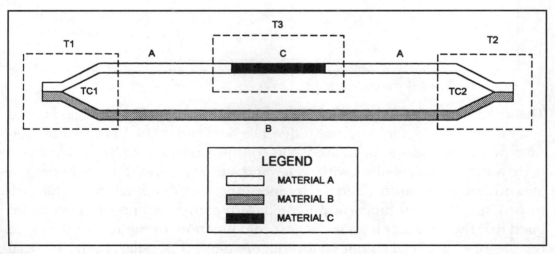

*Figure 6-7. Form of circuit for Law of Intermediate Materials.*

The properties of a thermocouple of materials A and C can be deduced from observation of the respective behaviors of these materials against a third, reference, material (B). The reference material selected is usually platinum or palladium. If potentials V1 and V2 are the thermoelectric potentials of A-B and B-C, respectively, then V1-V2 is the total system potential of the two thermocouples together.

## Thermocouple Output Voltage

The differential voltage between the two thermocouple junctions is proportional to the temperature difference, and is used as the output voltage. This potential is found from the equation:

$$V = a + bT + cT^2 + dT^3 + eT^4 + fT^5 \qquad \text{eq. (6-17)}$$

Where:

V is the output potential in volts.

T is the temperature of the measurement junction.

a, b, c, d, e and f are constants that are a function of the materials used in the thermocouple.

In many practical cases, only the first three terms may be used (making the equation a quadratic), while in others the first four terms are used.

## Linearizing a Thermocouple

The equation governing the thermocouple demonstrates a strong nonlinear dependence of the output voltage on temperature. As explained above, in some cases an approximation of the output voltage is made using just the quadratic version of the equation (cubic and higher terms deleted or approximated with an additional constant). This practice was especially reasonable when better linearization methods were not easily available. Analog circuits to solve the quadratic equation are, after all, somewhat easier than circuits for the cubic equation. But today there is the possibility of using a computer linearization method.

As with other systems, there are more than one way to linearize the thermocouple. For example, a circuit called a diode breakpoint generator can be used to piecewise linearize a circuit. These circuits use a series of diodes, each biased to a different point, in the gain setting network of an amplifier. When a diode is turned on by an input voltage that overcomes the bias, the gain of the amplifier is altered. The gain change linearizes the input circuit. But that system is both cumbersome and subject to, of all things, thermal drift in the breakpoint generator diode circuits (this same phenomenon forms the basis for our next category of sensors - semiconductors).

It is also possible to use a computer or computer-like circuit for linearization. The two computer methods involve: a) a look-up table to correct the value of output voltage for any given temperature, and b) an algorithm that will solve Equation 6-18 for T at any given output voltage. In both cases, the computer can be programmed with information on the specific type of thermocouple being used so that either the correct look-up table or the correct values of the coefficients of Equation 6-18 are selected from a standard table.

$$ T = a_o + a_1V + a_2V^2 + a_3V^3 + ... + a_nV^n \qquad eq. (6\text{-}18) $$

## Construction of Practical Thermocouple Sensors

The basic thermocouple is made by fusing together two or more wires of differing properties in order to form a junction. In some cases, this simple junction is used. In other cases, however, the thermocouple must be packaged. Figure 6-8 shows three basic forms of thermocouple package. In Figure 6-8a is the exposed tip thermocouple. The wires of the thermocouple are passed through (and insulated by a manganese oxide sheath from) a metallic tube (which provides mechanical protection and electrical shielding), but are exposed to the environment at the tip. In the grounded thermocouple of Figure 6-8b the wires are insulated from the metal tube except at the end. Finally, in Figure 6-8c we see the insulated or floating thermocouple. The construction is similar to Figure 6-8b, except that the thermocouple wires are not attached to the metal shield. In this example, the insulating material fills the entire tube housing.

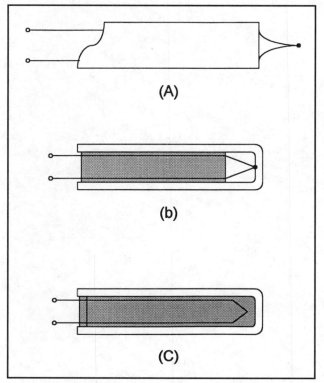

*Figure 6-8. Thermocouple packaging types.*

# Chapter 7

# Electro-Optical Sensors

Electro-optical sensors are electronic components that respond in some way or another to either light or the other electromagnetic waves in the infrared, ultraviolet and X-ray bands close to the visible light band. Photosensors that are easily available include photoemissive cells, photoresistors, photovoltaic cells, photodiodes and phototransistors. Below we take a look at each of these types. But first, by way of review, let's take a quick look at light.

## Light

Light is a form of electromagnetic radiation, and is thus essentially the same as radio waves, infrared (heat) waves, ultraviolet, and X-rays. The principal difference between these various types of electromagnetic waves is the frequency ($f$) and wavelength ($\gamma$) see Figure 7-1. The wavelength of visible light is 400 to 800 nanometers (1 nm = $10^{-9}$ meters), which roughly corresponds to frequencies between $7.5 \times 10^{14}$ and $3.75 \times 10^{14}$ Hz. Infrared radiation has wavelengths longer than 800 nm, and ultraviolet has wavelengths shorter than 400 nm; X-radiation has wavelengths even shorter than ultraviolet. We know that frequency and wavelength in electromagnetic waves are related by the equation:

$$\lambda = \frac{c}{f}$$

eq. (7-1)

Where:

      c is the velocity of light (300,000,000 meters/second).

      $\gamma$ is wavelength in meters.

      $f$ is frequency in hertz.

From the above equation you can see that light has a frequency on the order of $10^{14}$ hertz (compare with the frequencies of AM ($10^6$ Hz) and FM ($10^8$ Hz) broadcast bands in the radio portion of the spectrum).

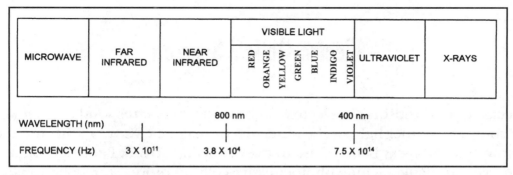

*Figure 7-1. Electromagnetic spectrum in vicinity of visible light.*

Because IR, UV and X-radiation are similar in both basic nature and wavelength to visible light, many of the sensors and techniques applied to visible light also work to one extent or another in those adjacent regions of the electromagnetic spectrum. Although performance varies somewhat, and some devices aren't useful at all in certain spectral regions, it is nonetheless true that people whose applications deal with those spectra may find these devices useful.

The photosensors described in this chapter depend upon quantum effects for their operation. Quantum mechanics arose as a new idea in physics in December 1900, the very dawn of the twentieth century, with a now-famous paper by German physicist named Max Planck. He had been working on thermodynamics problems ("black body" radiation), and found the experimental results reported in 19th-century physics laboratories could not be explained by classical Newtonian mechanics; the then-prevailing world view of physics. The solution to the problem turned out to be a simple, but terribly revolutionary idea: energy exists in discrete bundles, not as a continuum.

In other words, energy comes in packets of specific energy levels; other energy levels are excluded. The name eventually given to these energy levels was quanta. The name given to energy bundles that operated in the visible light range was photons.

The energy level of each photon is expressed by the equation:

$$E = \frac{c\,h}{\lambda}$$  eq. (7-2)

or alternatively,

$$E = h\,v$$  eq. (7-3)

Where:

E is the energy in electron volts (eV).

c is the velocity of light ($3 \times 108$ m/s).

$\gamma$ is the wavelength in meters (m).

h is Planck's constant ($6.62 \times 10^{-34}$ J-s).

v is the frequency of light in hertz (Hz).

The constant *ch* is sometimes combined, and expressed as 1240 eV/nm. The basis of light sensors is to construct a device that allows at least one electron to be freed from its associated atom by one photon of light. Materials in which the electrons are too tightly bound for light photons to do this work will not work well as light sensors.

Figure 7-2 shows the effect of prisms and filters on a light beam. White light contains the entire visible light spectrum from red to violet. When the white light is passed through a prism, long wavelengths (e.g. red light) are refracted less than short wavelengths (e.g. violet). As a result, the spectrum is spread out and because of the divergence you can see the individual colors that make up the white light.

Optical filters are materials that will selectively block or pass specific wavelengths. For example, if a green filter is placed in the path of a light beam, only green light will pass through the filter. This principle is used in a lot of

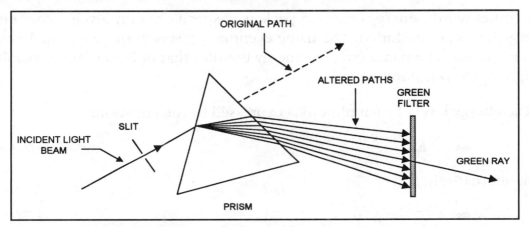

*Figure 7-2. Prism spreads the spectrum of white light (a phenomenon discovered by Sir Isaac Newton).*

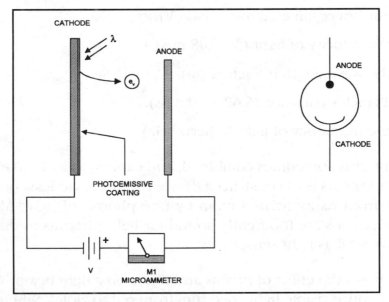

*Figure 7-3. Photoelectric effect.*

electronic instruments that find application in science, medicine and industry. Typical filters include colored glass, plastic, theatrical gels and other materials. For infrared, one can use developed color photographic film.

The spectral response of a sensor is a measure of its ability to respond to electromagnetic radiation of different wavelengths. The spectral response of a sensor is generally given in the form of a graph relating relative response to wavelength.

## Light Sensors

There are a number of solid-state light sensors used in electronics. These include: Photoemissive sensors (photoelectric and photomultiplier tubes), photoresistors, photovoltaic cells, photodiodes and phototransistors.

## Photoemissive Sensors

Photoemissive sensors are specially constructed vacuum tube diodes (i.e. "two electrode" devices) that output a current, $I_o$, that is proportional to the intensity of a light source that impinges its sensitive surface. Photoemissive sensors fall into two categories, photoelectric tubes and photomultiplier tubes. These devices depend on the photoelectric effect for their operation.

The photoelectric effect, the photoemission of electrons, was discovered in the early to mid-19th century. Scientists working on primitive batteries noticed that they worked more efficiently when exposed to light. In other experiments, it was found that certain metallic plates in a vacuum were able to emit electrons when exposed to visible or ultraviolet light. As early as 1887, radio research pioneer Heinrich Hertz was using spark-gap receiving apparatus in his experiments. He noted that the spark-gap receiver was more sensitive when the gap region was illuminated by ultraviolet light. William Hallwachs explained this phenomenon in 1888 by describing the photoelectric effect, also sometimes called the Hallwachs effect. The cause of the photoelectric effect, however, eluded scientists until early in this century.

Planck's 1900 paper that dealt with black body radiation did not address the photoelectric effect. It was, however, the prodromal work that led Albert Einstein to the solution to the photoelectric problem. In 1905, truly a seminal year in physics, Einstein published three major papers in Annalen der Physik. Included were the Theory of Relativity, a paper that explained Brownian motion, and a theory of the photoelectric effect that used Planck's energy quanta. It was for his work on the photoelectric effect that Albert Einstein was awarded the Nobel Prize, not for relativity as is commonly assumed.

The photoelectric effect perplexed scientists up to 1905 because of a certain strange behavior observed in the laboratory. The intensity of the impinging light beam affects only the amount of current (i.e. number of electrons) emitted, but not the energy of the emitted electrons. Oddly, however, it was noted that the color of the light affects the energy of the electrons. Electrons emitted under the influence of blue light from a photoemissive surface were more energetic than electrons emitted under a red light. The explanation of this behavior can be inferred from Equations 7-2 and 7-3. The energy expression for the photoelectric effect is:

$$\frac{1}{2}mV^2 = h\nu - E_\omega \qquad\qquad \textit{eq. (7-4)}$$

Where:

m is the mass of the electron.

V is the velocity of the fastest emitted electrons.

h is Planck's constant ($6.63 \times 10^{-27}$ erg-second).

v is the frequency of the incident light.

$E_\omega$ is the energy in ergs required to permit an electron to escape from the photoemissive surface.

The photoelectric effect is not seen at all wavelengths. For an electron to be emitted, the applied photon energy must be greater than the work function energy ($E_\omega$) of the material. The work function is the amount of energy required to dislodge an electron from its associated atom. We may conclude, therefore, for any given material or metal, there is a wavelength limitation. Below a certain critical wavelength the photon energy is less than the work function energy. The maximum wavelength can be inferred from:

$$\lambda \leq \frac{c\,h}{E_\omega} \qquad\qquad \textit{eq. (7-5)}$$

Where:

λ is the wavelength in meters.

c is the speed of light ($3 \times 10^8$ m/s).

h is Planck's constant ($6.62 \times 10^{-34}$ J-s).

$E_\omega$ is the work function energy for the material illuminated.

The construction of a typical photoemissive sensor tube is shown in Figure 7-3. The device is a diode, so has two electrodes inside of an evacuated glass or metal housing. The photoemissive surface is called the cathode, while the electron collector electrode is called the anode. These elements are housed inside the glass or metal envelope, which is either under a high vacuum or filled with a partial pressure of an ionizing inert gas.

The anode is constructed of a small loop of wire or a thin rod, usually made of either tungsten, platinum or an alloy. The cathode is made of metal, but the actual surface is enhanced for improved emissivity by a special light sensitive coating. Typical materials used include antimony, silver, cesium, or bismuth. The material is usually mixed with trace quantities of other elements.

Although photoemission does not require an external voltage source, the electrons must be collected and delivered to an external circuit before the device is useful. In order to collect electrons, an electrical potential (V in Figure 7-3) is connected across the tube such that the anode is positive with respect to the cathode. Electrons, being negatively charged, are repelled from the negative cathode and are attracted to the positive anode. For most common photoemissive sensors $0 \leq V \leq 300$ volts.

There are two general forms of tube construction. In side-excited designs (Figure 7-4a) the cathode is constructed such that the light must enter the side wall of the tube. The other directions may be blinded by internal blackening or silvering in order to prevent stray light from exciting the photoemissive surface. The end-excited design is constructed with the photoemissive surface facing the end of the housing (Figure 7-4b). Again, all other surfaces than the optical window may be blinded against stray light.

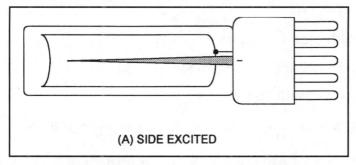

Figure 7-4a. Photoemissive tubes: Side excited.

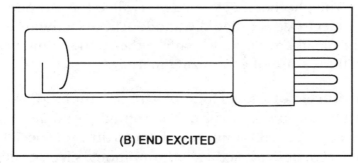

Figure 7-4b. Photoemissive tubes: End excited.

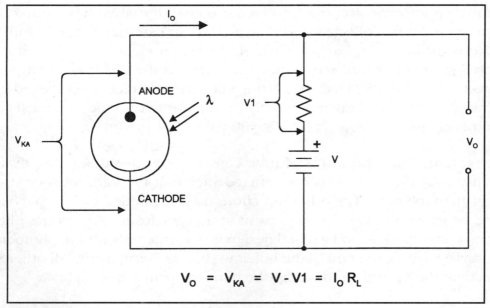

$$V_O = V_{KA} = V - V1 = I_O R_L$$

Figure 7-5. Photoemissive tube circuit.

Photoemissive sensors are also categorized according to whether or not the inside is maintained under a vacuum or is gas-filled. A high vacuum photodetector is evacuated of all air and gases. The device outputs a small current $(I_o)$ that is linearly proportional to the intensity of the impinging light.

The output current levels from the high vacuum photodetector are generally too low to be measured directly on a meter movement, so they are typically converted to an output voltage $(V_o)$ by passing the current through a load resistor ($R_L$ in Figure 7-5).

The output voltage, $V_o$, is the difference between the power supply potential, $V$, and the voltage drop (V1) across $R_L$ caused by the photodetector output current, $I_o$. Because, by Ohm's law, V1 is equal to the product $I_o R_L$, we know that the output voltage is:

$$\bullet \quad V_o = V - I_o R_L \qquad\qquad eq.\ (7\text{-}6)$$

A gas-filled photoemissive sensor is first evacuated of air, and then refilled with an ionizing inert gas. Photoelectrons emitted by the cathode collide with the gas molecules, causing the creation of electrons and positive ions by secondary emission. Thus, each of the emitted electrons creates a number of secondary electrons, so the overall current flow is ten to nearly one-thousand times greater than in a similar high vacuum device. The gas pressure is carefully regulated to ensure that this process happens without either snuffing out or running away. In addition, the cathode to anode potential $(V_{ka})$ must be kept low enough to prevent imparting sufficient kinetic energy to the electrons to cause complete ionization of the gas. If this were to happen, then the phototube would emit light in the same manner as a glow lamp. The gas filled tube produces higher output current than the high vacuum types, but does not have as linear a light intensity vs. output current characteristic.

Photoemissive sensors have a specification called the dark current, $I_d$, i.e. a current flowing from cathode to anode when there is no light impinging the photoemissive surface. For gas-filled tubes, $I_d$ is on the order of $10^{-7}$ to $10^{-8}$ amperes, while for high vacuum devices it is $10^{-8}$ to $10^{-9}$ amperes.

# Electronic Circuit Guidebook, Volume 1: Sensors

The response time of the photoemissive sensor is a measure of the time required for the device to respond to changes in applied light level. For high vacuum devices, this time is on the order of one nanosecond (1 ns), while for gas-filled tubes it is one millisecond (1 ms).

## Photomultiplier Tubes

The photoemission process is less efficient than is needed in many cases, especially under low light-level conditions. The photoelectric tube system can be made more efficient by using a photomultiplier tube (Figure 7-6).

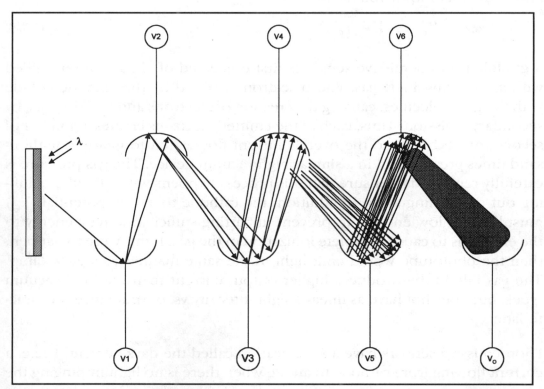

*Figure 7-6. Structure of photomultiplier tube.*

In this type of photosensor there are a number of positively charged anodes, called dynodes, that intercept the electrons. When light impinges on the photoemissive cathode, electrons are emitted due to the photoelectric effect. They are accelerated through a positive high voltage potential (V1) to the first dynode. They acquire substantial kinetic energy during this transition, so when each electron strikes the metal it gives up its kinetic energy. In the giving up of this kinetic energy, some is converted to heat, while some kinetic energy is converted by dislodging additional "secondary emission" electrons from the dynode surface. Thus, a single electron caused two or more additional electrons to be dislodged. These electrons are accelerated by high voltage potential V2, and reproduce the same effect at the second dynode. The process is repeated several times (each voltage step being about 75 to 100 higher times than the preceding step), and each time several more electrons join the cascade for each previously accelerated electron. Finally, the electron stream is collected by the last dynode or a separate anode, and can be used in an external circuit.

Figure 7-7 shows how a photomultiplier might be mounted in a light-tight housing that admits light only through an aperture and (possibly) a filter.

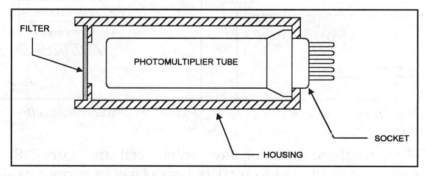

*Figure 7-7. Packaging for photomultiplier tube.*

## Photovoltaic Cells

A photovoltaic (also called photogalvanic) cell is a device in which an electrical potential difference is generated, and thus a current made to flow in an external circuit, by shining light onto its surface. The common "solar cell" is an example of the photovoltaic cell.

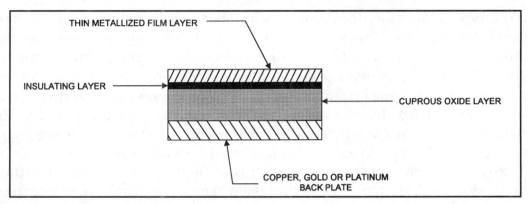

*Figure 7-8a. Photovoltaic cells: Copper oxide photovoltaic cell.*

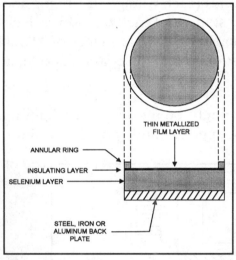

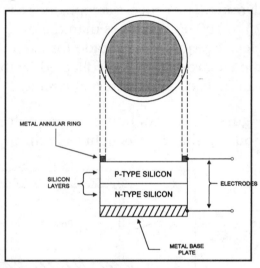

*Figure 7-8b. Photovoltaic cells:*
*Selenium cell.*

*Figure 7-8c. Photovoltaic cells:*
*Silicon cell.*

Figure 7-8 shows three forms of photovoltaic cell. In Figure 7-8a, a metal disk (copper, gold or platinum) is coated with layer of copper oxide, which is in turn covered with a semi-transparent layer that passes light and collects emitted electrons. The copper-oxide cell was invented prior to World War I by Bruno Lange, and eventually marketed by Westinghouse under the tradename Photox cell.

A similar photovoltaic cell is made of selenium (Figure 7-8b). These cells were invented in the 1930s, and marketed by Weston Instruments under the trade name Photronic cell. In selenium cells, a layer of photosensitive selenium is coated onto an iron, steel or aluminum plate. In both forms of metal photovoltaic cell the thin insulator forms a barrier layer. When light illuminates the barrier layer, the impinging light photons are absorbed, and in the process electrons are emitted. The existence of free electrons causes a difference of electrical potential to appear across the barrier layer. The selenium cell is negative, while the transparent thin metal film side is positive.

Selenium cells produce an output range of 0.2 to to 0.6 volts dc, and 0.45 volts dc under 2,000 foot-candles of illumination. Photovoltaic cells designed for power applications produce between 20 and 90 milliwatts (mw) of dc power per square inch of photoactive surface exposed to light. The selenium cell covers a spectrum of 300 to 700 nm, with its spectral response being centered near 560 nm. Like other sensors, it is common practice to alter the response of a selenium cell with special filters over the transparent window. For most instrumentation purposes, the selenium cell must be loaded with a resistor. Otherwise, the characteristic curve is highly nonlinear.

Figure 7-8c shows the structure of the silicon photovoltaic cell, discovered in 1958 by scientists at Bell Telephone Laboratories. The silicon cell consists of a PN junction of P and N-type silicon. In the form shown in Figure 7-8c, the N-type silicon is deposited onto a metallic substrate, that also forms the negative terminal of the cell. The P-type layer is diffused over the N-type, and forms the surface exposed to light. The positive electrode is an annular ring deposited onto the exposed surface of the P-type silicon region. These cells output a potential of 0.27 to 0.6 volts under illumination of 2,000 foot-candles.

Figure 7-9 shows a typical circuit for instrumentation applications of the photovoltaic cell. The cell is connected across the input of a high impedance amplifier, such as the noninverting operational amplifier shown. The output voltage is found from the product of the op-amp gain and photocell output voltage. The operational amplifier provides a high input impedance buffer between the cell and the "outside world."

## Photoconductive Cells

A photoconductive cell (photoresistor) is a device that changes electrical ohmic resistance when light is applied. The usual circuit symbol for photoresistors is the normal resistor symbol enclosed within a circle, and given the greek lambda symbol to denote that it is a resistor that responds to light.

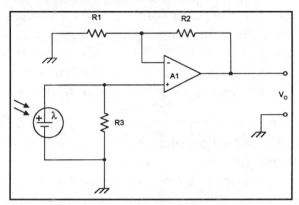

*Figure 7-9. Circuit for interfacing photovoltaic cell.*

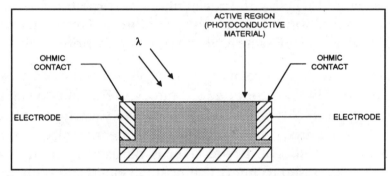

*Figure 7-10. Photoconductive or photoresistor cell.*

The active region (see Figure 7-10) of a photoconductive cell is a thin film of silicon, germanium, selenium, a metallic halide, or a metallic sulfide (e.g. cadmium sulfide, CdS). When these types of materials are illuminated with light, free electrons are created as photon energy drives them from a valence band into the conduction band. As in any conductor, free electrons mean that current can flow if an electrical potential is applied. Because increased illumination creates additional free electrons which are available for conduction, the resistance of the photoconductive cell decreases with increases in illumination level.

When photoresistors are specified, it is typical that a dark resistance is given, as is a light/dark ratio. Changes of megohms when dark, to hundreds of ohms under maximum illumination, are common. In most common varieties, the resistance is very high when dark, and drops very low under intense light. The intensity of the light affects the resistance, so they can be used in photographic lightmeters, densitometers, colorimeters and so forth.

# Chapter 8

# Magnetic Sensors

Magnetic fields are of significant interest, so it is no surprise that a number of magnetic sensors are available. In this chapter we will look at Hall effect devices and flux gate sensors, plus some applications such as magnetometers and gradiometers.

Magnetometers are used in a variety of applications in science and engineering. One high tech magnetometer is used by Navy aircraft to locate submarines. Radio scientists use magnetometers to monitor solar activity. Metal detectorists and archeologists use magnetometers to locate buried treasure, marine archeologists and treasure hunters use the devices to locate sunken wrecks and sunken treasure.

Gradiometers are differential magnetometers, i.e. a system of two balancing magnetometers, usually in the vertical plane, that produce equal outputs when there are no magnetic anomalies in the vicinity of either sensor.

## Hall Effect Devices

The Hall effect device is one of the oldest (1879) known sensors, although it has only been available in practical form since the 1950s. The Hall effect is named after its discoverer, physicist E.W. Hall. Unfortunately, at the time Hall discovered the effect it was only usable as a laboratory curiosity because suitable materials were not yet available, at least in commercially usable form and quantities. In the 1950s, the Hall effect sensor became a possibility because of the development of indium arsenide.

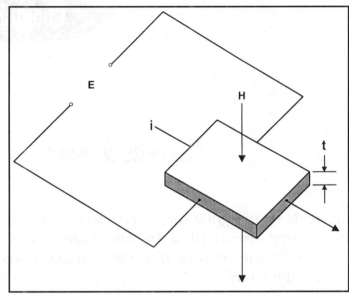

*Figure 8-1. Hall effect magnetic sensor.*

Figure 8-1 shows the basic form of the Hall effect sensor. A slab of Hall material, of thickness t, has a current i passing through it, and a magnetic field H perpendicular to it. An electrical potential, i.e. a voltage, is developed across the axis that is perpendicular to both the current flow and the magnetic field. The Hall potential is:

$$E = \frac{R H i}{t}$$

*eq. (8-1)*

Where:

E is the Hall potential in volts (V).

R is the Hall constant for the material.

H is the magnetic field.

i is the control current.

t is the device thickness.

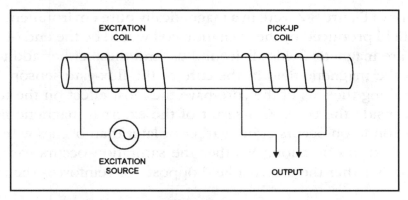

*Figure 8-2. Simple form of flux-gate magnetometer.*

The value of $E$ can be quite different for different materials. Values on the order of 10-15 mA/1,000 gauss to 100 mA/1,000 gauss are found. However, the output tends to be linear up to 10 gauss or so. Also, the device tends to be a bit temperature sensitive, which is one of its two major drawbacks. Compensation circuitry is needed to overcome a temperature coefficient of about one-tenth percent per degree Celsius (0.1%/°C).

The other problem with Hall sensors is a certain frailtydue to the material. Hall sensors can be adversely affected by over-temperature, too much excitation current, or mechanical shock.

## Flux-Gate Sensors

Flux-gate magnetic sensors are basically over-driven magnetic core transformers in which the transducible event is the saturation of the magnetic material. These devices can be made very small and compact, yet provide reasonable accuracy.

Figure 8-2 shows the most simple form of flux-gate magnetometer sensor. It consists of a nickel-iron rod used as a core, wound with two coils. One coil is used as the excitation coil, while the other is used as the output or sensing coil. The excitation core is driven with a square wave (Figure 8-3) with an amplitude high enough to saturate the core. The output current signal will increase in a linear manner so long as the core is not saturated. But when the saturation point is reached, the inductance of the coil drops and the current rises to a level limited only by the other coil circuit resistances.

If the sensor of Figure 8-2 were in a magnetically pure environment, then the magnetic field produced by the excitation coil would be the end of the story. But there are magnetic fields all around us, and these either add to or subtract from the magnetic field in the core of the flux-gate sensor. Magnetic field lines along the axis of the core have the most effect on the total magnetic field inside the core. As a result of the external magnetic fields, the saturation condition occurs either earlier or later than occurs with only the excitation field in operation. Whether the saturation occurs early or later depends on whether the external field opposes or reinforces the excitation field.

The variation in entering saturation state is the transducible event on which this type of sensor is based. Unfortunately, it's also a bit difficult to recover this information. A better solution is shown in Figure 8-4. In this version of the flux-gate magnetometer there are two independent cores, each of which has its own excitation winding. A common pick-up winding serves both cores. The excitation coils are wound in the series-opposing manner such that the inductions in the cores precisely cancel each other if the external field is zero. The external field causes pulses to arise in the pick-up coil that can be integrated in a low-pass filter to produce a slowly varying dc signal that is proportional to the applied external magnetic field.

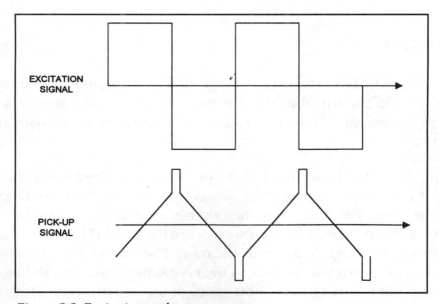

*Figure 8-3. Excitation and output currents.*

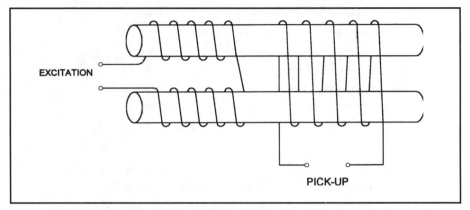

*Figure 8-4. Dual core magnetometer is an improvement on the basic design.*

## Toroidal Core Flux-Gate Sensor

The straight core flux-gate sensors suffer from two main problems. First, the desired signal is small compared with the signal on which it rides, so is difficult to discriminate properly. Second, there must be a very good match between the cores and the excitation winding segments on each winding. While these can be overcome, it becomes expensive and thus suffers in popularity.

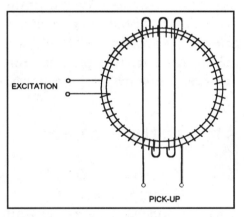

*Figure 8-5. Toroid core flux-gate magnetometer.*

A better solution is to use a toroidal or "doughnut" shaped magnetic core (Figure 8-5). This type of core relieves the problem of picking off small signals in the presence of large offset components. It also reduces the drive levels required from the excitation source.

In the toroidal core flux-gate sensor we can get away with using a single excitation coil wound over the entire circumference of the toroidal core. The pick-up coil is wound over the outside diameter of the core, rather than around the ring as is the excitation coil.

Another advantage of the toroidal core form of magnetometer sensor is that a pair of orthogonal (i.e. right angle) pick-up cores can be installed that will allow null measurements to be made.

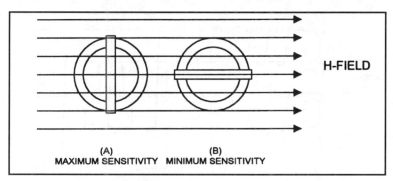

*Figure 8-6. Orientation effects on sensitivity.*

Figure 8-6 shows the orientation of the toroid core flux-gate sensor as a function of sensitivity. The maximum sensitivity occurs when the magnetic H-field is orthogonal to the pick-up coil, while minimum sensitivity occurs when the pick-up coil and H-field are aligned with each other. As you can see, when there are two pick-up coils at right angles to each other, then one will be most sensitive as the other goes through null condition.

## A Practical Flux-Gate Sensor

A compact and reasonably low cost line of flux-gate sensors, designated FGM-x, is made by Speake & Co. Ltd (Elvicta Estate, Crickhowell, Powys, Wales, UK), and distributed in the United States by Fat Quarters Software: 24774 Shoshonee Drive, Murrieta, CA 92562; 909-698-7950 (voice) and 909-698-7913 (FAX). The FGM-3 device is the one that I experimented with when preparing this article. It is 62 mm long by 16 mm diameter (2.44" × 0.63"). These devices convert the magnetic field strength to a signal with a proportional frequency. One of the things I found fascinating about the FGM-3 is that a set of only three leads provides operation:

Red    +5 Vdc (Power)

Black    0 Volts (Ground)

White    Output Signal (Square Frequency)

The magnetic detection rating of the device is ±0.5 Oersted (±50 μTesla). This range covers the earth's magnetic field, making it possible to use the sensor in earth field magnetometers. Using two or three sensors in conjunction with each other provides functions such as compass orientation, three-dimensional orientation measurement systems and three-dimensional gimballed devices such as virtual reality helmet display devices. It can also be used to provide magnetometry (including Earth field magnetometry), ferrous metal detectors, underwater shipwreck finders, and in factories as conveyer belt sensors or counters. There are a host of other applications where a small change in magnetic field is the important transduction event.

The packages for the FGM-1 and FGM-3 sensors are shown in Figure 8-7. The FGM-3 has the three leads discussed above, and is of the 62 mm x 16 mm size. The FGM-1 device is smaller than the FGM-3, being 30 mm (1.18") long by 8 mm (0.315") diameter. It has a small connector on one end consisting of four pins: 1) feedback; 2) signal output; 3) ground, and 4) +5 Vdc power. The signal, output and ground terminals are essentially the same as on the FGM-3, but the feedback pin provides some extra flexibility. The feedback pin leads to an internal coil that is wound over the flux-gate sensor. It is used to alter the zero field output frequency, or to improve linearity of the sensor over the entire range of the sensor.

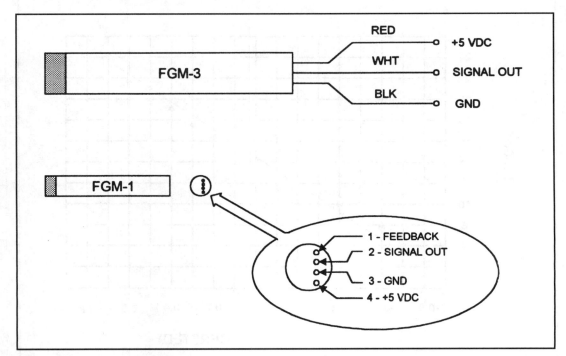

*Figure 8-7. Speake & Co. Ltd. FGM-1 and FGM-3 sensors.*

# Electronic Circuit Guidebook, Volume 1: Sensors

The FGM-series device output is a +5 volt (TTL-compatible) pulse whose period is directly proportional to the applied magnetic field strength. This relationship makes the frequency of the output signal directly proportional to the magnetic field strength. The period varies typically from 5.5 μS to 25 μS (Figure 8-8), or a frequency of about 120 Khz to 50 Khz. For the FGM-3 the linearity is about 5.5 percent over the ±0.5 Oersted range.

The FGM-1, FGM-2 and FGM-3h sensors are related to the FGM-3. The FGM-1 is the smaller version of the FGM-3, with a range of ±0.7 Oersted (±70 μTesla). The FGM-2 is an orthogonal sensor that has two FGM-1 devices on a circular platform at right angles to one another. This orthogonal arrangement permits easier implementation of orientation measurement, compass and other applications. The FGM-3h is the same size and shape as the FGM-3, but is about 2.5 times more sensitive. The output frequency changes approximately 2 to 3 Hz per gamma of field change, with a dynamic range of ±0.15 Oersted (about one-third the earth's magnetic field strength).

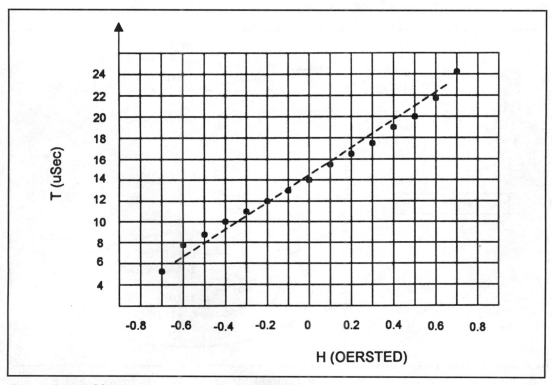

*Figure 8-8. Calibration points on a typical FGM-3.*

The response pattern of the FGM-x series sensors is shown in Figure 8-9. It is a "figure-8" pattern that has major lobes (maxima) along the axis of the sensor, and nulls (minima) at right angles to the sensor axis. This pattern suggests that for any given situation there is a preferred direction for sensor alignment. The long axis of the sensor should be pointed towards the target source. When calibrating or aligning sensor circuits it is common practice to align the sensor along the east-west direction in order to minimize the effects of the earth's magnetic field.

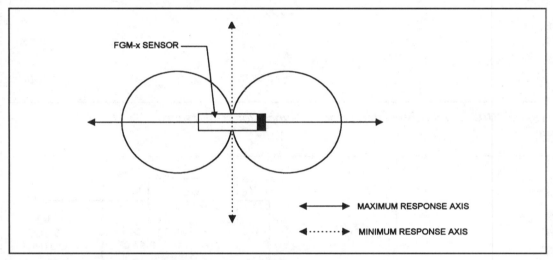

*Figure 8-9. Sensitivity pattern for the FMG-x devices.*

## Powering the Sensors

The FGM-x series of flux-gate magnetic sensors operates from +5 volts dc, so is compatible with a wide variety of analog and digital support circuitry. As is usual for any sensor, you will want to use only a regulated dc power supply for the FGM-x devices. In fact, the manufacturer recommends that double-regulation (Figure 8-10a) be used. Ripple in the dc power supply can cause output frequency anomalies, and those should be avoided. In the circuit of Figure 8-10a, an unregulated +12 to +15 Vdc input potential is applied first to a 9 volt 78L09 or 78M09 three-terminal IC voltage regulator (U1). This produces a +9 volt regulated potential that is then applied to the input of the 78L05 or 78M05 device (U2). The second regulator reduces the +9 volts from U1 to the +5 volts needed for the FGM-x sensors.

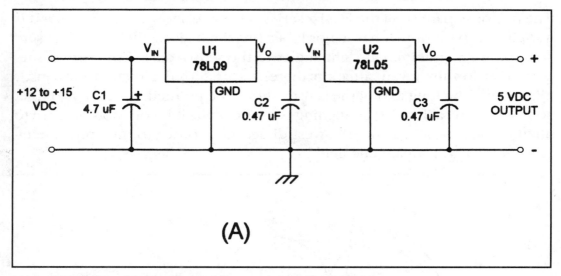

Figure 8-10a.  Double regulated dc power supply for FGM-x sensors.

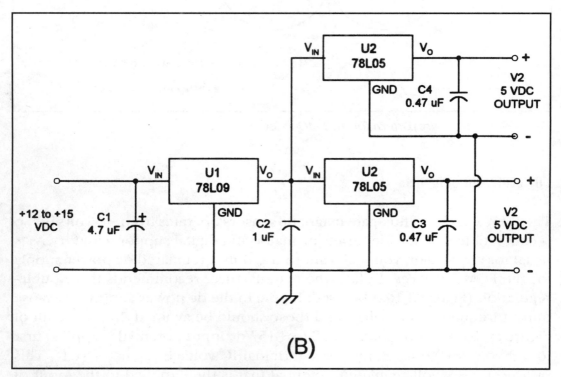

Figure 8-10b.  Supply to use when other circuits must be powered.

When other digital devices are being powered from the same dc power supply it is prudent to provide a separate dc source for the FMG-x sensors. In Figure 8-10b we see the type of circuit that would accomplish this task. There are two separate +5 Vdc outputs, labelled V1 and V2. Both are derived from 78L05 devices that are powered from a single 78L09. Care must be taken to not exceed the maximum current limits of U1, especially if the same size IC voltage regulators are used for all three (U1, U2, and U3). One of the +5 VDC sources, either V1 or V2, can be used for powering the FGM-x device, while the other powers the rest of the circuitry.

## Calibration of the Sensors

The FGM-x devices are not precision instruments out of the box, but can be calibrated to a very good level of precision. The calibration chore requires you to generate a precise magnetic field in which the sensor can be placed. One way to generate well-

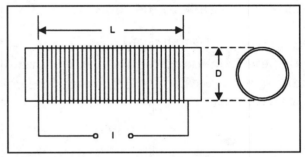

Figure 8-11. Solenoid-wound calibration coil.

controlled and easily measured magnetic fields is to build a coil and pass a dc current through it. If the sensor is placed at the center of the coil (inside), then the magnetic field can be determined from the coil geometry, the number or turns of wire and the current through the coil. There are basically two forms of calibrating coil found in the various magnetic sensor manuals: solenoid-wound and the Helmholtz pair.

Figure 8-11 shows the solenoid coil. A solenoid is a coil that is wound on a cylindrical form in which the length of the coil (L) is greater than or equal to its diameter. This type of coil is familiar to radio fans because it is used in many L-C tuning circuits. The magnetic field (H) in Oersteds is found from:

$$H = \left( \frac{4\pi N I L}{10 \ SQR(L^2 + D^2)} \right)$$

eq. (8-2)

Where:

H is the magnetic field in Oersteds.

N is the number of turns-per-centimeter (t/cm) in the winding.

I is the winding current in amperes.

D is the mean diameter of the winding in centimeters (cm).

The winding is usually made with either #24 or #26 enameled or Formvar® covered copper wire. The length of the solenoid coil should be at least twice as long as the sensor being calibrated, and the sensor should be placed as close as possible to the center of the long axis of the coil.

The Helmholtz coil is shown in Figure 8-12. It consists of two identical coils (L1 and L2) mounted on a form with a radius R, and a diameter 2R. The coils are spaced one radius (1R) apart. The equations for this type of calibration assembly are:

$$H = \frac{0.8991\,N\,I}{R} \qquad \text{eq. (8-3)}$$

and,

$$B = \frac{9.1 \times 10^3\,N\,I}{R} \qquad \text{eq. (8-4)}$$

In the practical case, one usually knows the dimensions of the coil, and needs to calculate the amount of current required to create a specified magnetic field. We can get this for the Helmholtz pair by rearranging Equation 8-3:

$$I = \frac{R\,H}{0.8991\,N} \qquad \text{eq. (8-5)}$$

The coils are a little difficult to wind, especially those of large diameter (e.g. 4"). One source recommends using double-sided tape (the double-sticky stuff) wrapped around the form where the coils are to be located. As the wires are laid down on the form they will stick to the tape, and not dither around.

The above equations, plus a lot of magnetic theory and calibration suggestions, plus information on other sensors, are found in Janicke (1994).

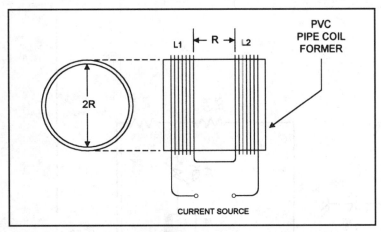

*Figure 8-12. Helmholtz pair calibration coil.*

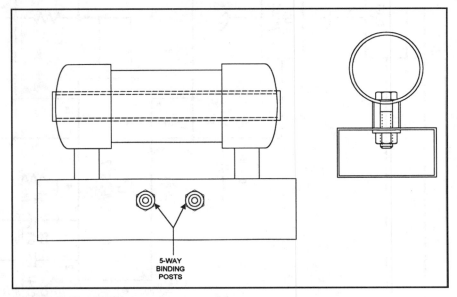

*Figure 8-13. Coil-mounting assembly.*

Figure 8-13 shows a type of assembly that can be used for either the solenoid or Helmholtz coil. I first saw this type of assembly in a college freshman physics laboratory about 25 years ago. It consists of a PVC pipe section used as the coil former. Endcaps on the coil former also serve as mountings. The mounts at either end consist of smaller segments of PVC pipe and nylon (non-magnetic) hardware fasteners. Another segment of PVC pipe, of much smaller diameter than the coil former, is passed through the former from

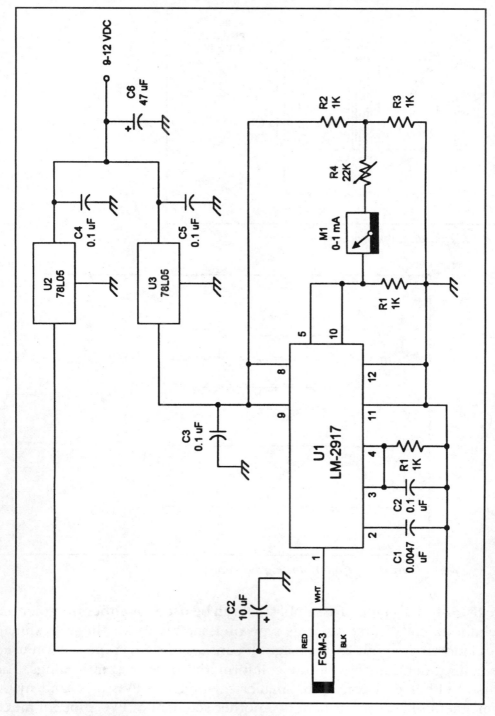

*Figure 8-14. Analog magnetometer using the LM-2917 F/V converter.*

one endcap to the other, such that its ends protrude to the outside. This pipe forms a channel into which the sensor can be placed. The base is a plastic or wooden box (again, non-magnetic materials). One thing nice about this type of assembly is that the sensor is always in approximately the same position in the coil, close to the center of the field.

## Analog Interface to FGM-3

Figure 8-14 shows a method for providing an analog interface to the FGM-3 and its relatives. The output of the sensor is a 40 to 125 kHz frequency that is proportional to the applied magnetic field. As a result, we can use a frequency-to-voltage (F/V) converter such as the LM-2917 to render the signal into a proportional dc voltage. That voltage, in turn, can be used to drive an analog or digital voltmeter or milliammeter. The LM-2917 is selected because it is widely available at low cost from mail order parts distributors.

The output circuit consists of a bridge made up of R1, R2 and R3, along with the output of the LM-2917 device. R2/R3 form a resistive voltage divider that produces a potential of 1/2-(V+) at one end of the 22K sensitivity control (R4). If the voltage produced by the LM-2917 is the same as the voltage at R4, then the differential voltage is zero and no current flows. But if the LM-2917 voltage is not equal, then a difference exists and current flows in R4/M1. That current is proportional to the applied magnetic field. Meter M1 and R4 can be replaced with a digital voltmeter, if desired.

The dc power supply uses two regulators, one for the FGM-3 and one for the LM-2917. Even better results can be obtained if an intermediate voltage regulator, say a 78L09, is placed between the V+ source and the inputs of U2 and U3. That results in double-regulation, and produces better operation.

## Digital "Heterodyning"

The method shown in Figure 8-15 results in a more sensitive measurement over a small range of the sensor's total capability. The circuit makes it possible to measure small fluctuations in a relatively large magnetic field.

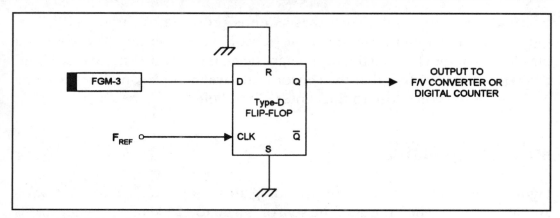

*Figure 8-15. Digital heterodyning ("digidyning") circuit.*

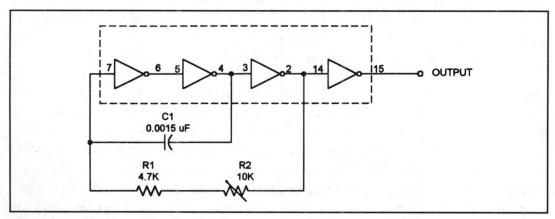

*Figure 8-16a. CMOS reference oscillator.*

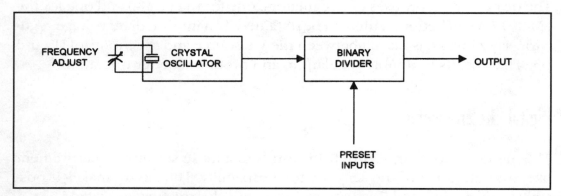

*Figure 8-16b. Crystal reference sourcer.*

A Type-D flip-flop is used in Figure 8-15 to "mix" the frequency from the FGM-3 with a reference frequency ($F_{REF}$). The FGM-3 literature calls this process "digital heterodyning" (dare we call it "digidyning"?), although it is quick to point out that it is really more like the production of alias frequencies by undersampling than true heterodyning.

Two types of frequency source can be used for $F_{REF}$. For relatively crude measurements, such as a passing vehicle detector, the CMOS oscillator of Figure 8-16a is suitable. This circuit is based on the 4049 hex inverter chip connected in an astable multivibrator configuration. The exact frequency can be adjusted using R2, a 10 kohm, 10-turn potentiometer.

Where a higher degree of stability is needed, for example when making earth field variation measurements or testing magnetic materials, a more stable frequency source is needed. In that case, use a circuit such as Figure 8-16b. This circuit uses a crystal controlled oscillator feeding a binary divider network. Crystal oscillators can be built, or if you check the catalogs you will find that a large number of frequencies are available in TTL and CMOS compatible formats at low enough costs to make you wonder why you would want to build your own. I've seen them sold for about the price of a crystal alone in catalogs such as Digi-Key.

The reference frequency is adjusted to a point about 500 Hz below the mean sensor frequency. This frequency is measured when the sensor is in the east-west direction. This arrangement will produce a frequency of 0 to 1,000 Hz over a magnetic field range of ±500 gamma.

## Magnetometer Project & Kit

Figure 8-17 shows the circuit for a simple magnetometer based on the FGM-3 flux-gate sensor. It can be obtained in kit form from Fat Quarters Software. The connections to the printed circuit board are shown in Figure 8-18. This device takes the output frequency of the FGM-3, passes it through a special interface chip (U1) and then to a digital-to-analog (D/A) converter to produce a voltage output.

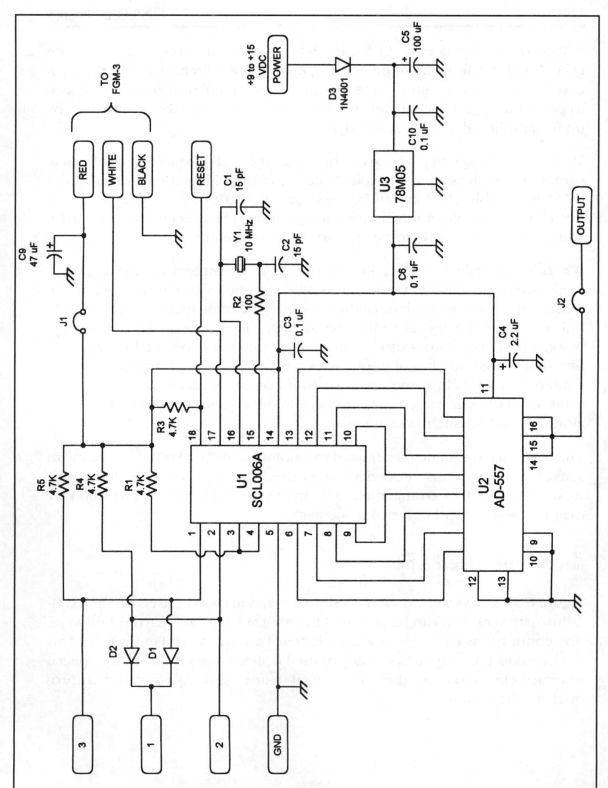

Figure 8-17. Magnetometer project circuit.

The heart of this circuit, other than the FGM-3 device, is the special interface chip, Speake's SCL006 device. It provides the circuitry needed to perform magnetometry, including earth field magnetometry. It integrates field flucuations in one-second intervals, producing very sensitive output variations in response to small field variations. It is of keen interest to people doing radio propagation studies, and who need to monitor for solar flares. It also works as a laboratory magnetometer for various purposes. The SCL006A is housed in an 18-pin DIP IC package.

The D/A converter (U2) is an Analog Devices type AD-557. It replaces an older Ferranti device seen in the Speake literature because that older device is no longer available. Indeed, being a European device it was a bit hard to

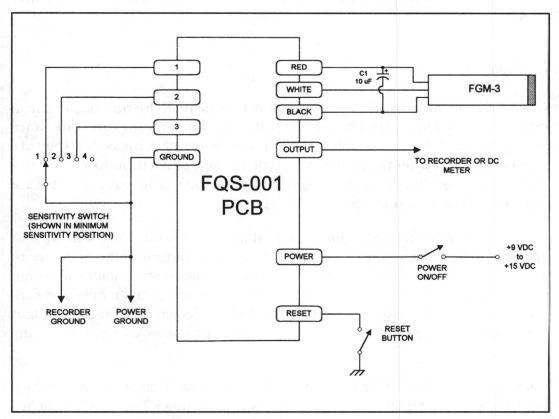

*Figure 8-18. PCB connections for the magnetometer.*

find in unit quantities required by hobbyists on this side of the Atlantic. The kit from Fat Quarters Software contains all the components needed, plus a printed circuit board. The FGM-3 device is bought separately.

The external connections are shown in Figure 8-18. The circuit is designed to be run from 9-volt batteries so that it can be used in the field.n A sensitivity switch provides four positions, each with a different overall sensitivity range. The output signal is a dc voltage that can be monitored by a strip-chart or X-Y paper recorder, voltmeter, or fed into a computer using an A/D converter.

If you intend to use a computer to receive the data, then it might be worth-while to eliminate the D/A converter and feed the digital lines (D0-D7) from the SCL006A directly to an eight-bit parallel port. Not all computers have that type of port, but there are plug-in boards available for PCs, as well as at least one product that makes an eight-bit I/O port out of the parallel printer port.

## Gradiometers

One of the problems with magnetometers is that small fluctuations occur in otherwise very large magnetic fields. And those fluctuations can sometimes be important. A further problem with single-sensor systems is that they are very sensitive to orientation. Even a small amount of rotation can cause un-acceptably large, but spurious, output changes. The changes are real, but are not the fluctuations that you are seeking.

A gradiometer is a magnetic instrument that uses two identical sensors that are aligned with each other so as to produce a zero output in the presence of a uniform magnetic field. If one of the sensors comes into contact with some sort of small magnetic anomaly, then it will upset the balance between sensors, producing an output. The gradiometer gets its name from the fact that it measures the gradient of the magnetic field over a small distance (typically 1 to 5 feet).

These instruments can be used for finding very small magnetic anomalies. For example, the metallic firing pin of plastic land mines buried a few inches below the surface, or a shipwreck buried deep in the ocean silt. Archeolo-

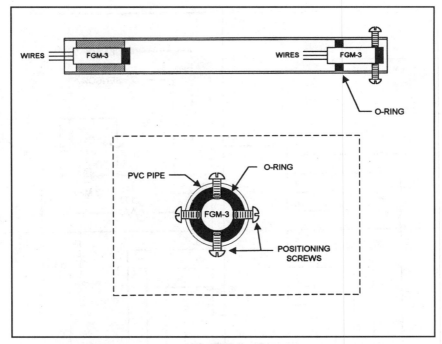

*Figure 8-19. Gradiometer instrument.*

gists use gradiometers to find artifacts, and identify sites. Also, people who explore Civil War battlefields, western mining camps, and other sites often use gradiometers to facilitate their work.

Figure 8-19 shows the construction details for a simple gradiometer based on the FGM-3 device. It is built using a length of PVC pipe. One sensor is permanently mounted at one end of the pipe, using any sort of appropriate non-magnetic packing material. In one experiment, I used the standard 0.5 inch adhesive backed window sealing tape used in colder areas of the country to keep the howling winds out of the house in wintertime. It worked nicely to hold the permanent sensor in place.

The other sensor is mounted in the opposite end of the tube using an O-ring that fits snugly into the tube. Four positioning screws made of non-magnetic materials are used to align the sensor. The position of the sensor is adjusted experimentally. The idea is to position the sensor such that the gradiometer can be rotated freely in space without causing an output variation.

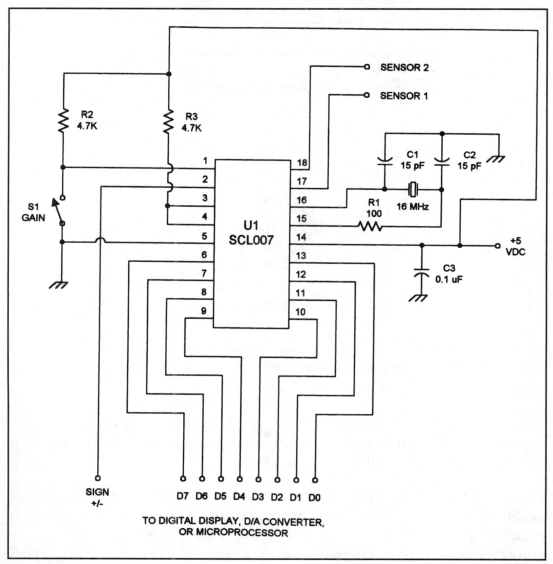

*Figure 8-20. Gradiometer circuit using the SCL007 device.*

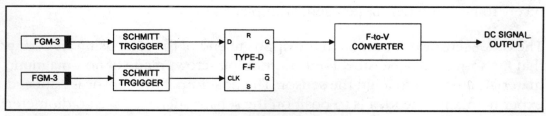

*Figure 8-21. Digidyning gradiometer.*

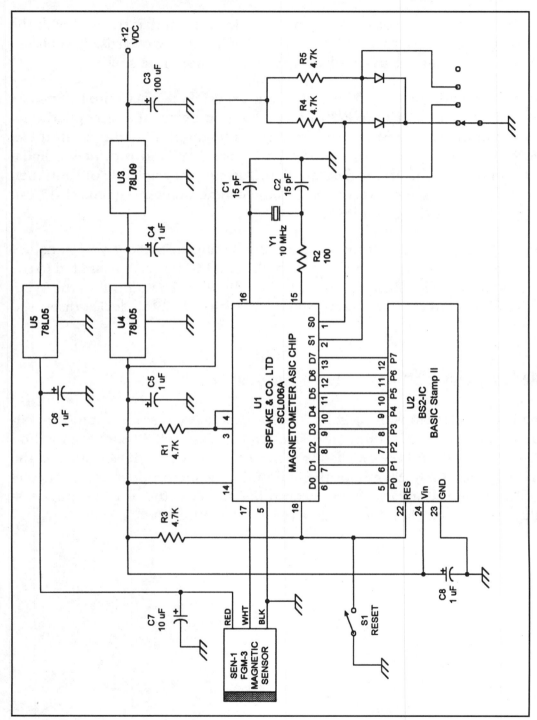

*Figure 8-22. Interfacing FGM-x series sensors to microcontrollers.*

The gradiometer sensor is usually held vertically such that the end with the wires coming out of the FGM-3 devices is pointed downwards. This allows you to find buried magnetic objects even if they are quite small.

A practical gradiometer can be built using a special interface chip by Speake, the SCL007 device (Figure 8-20). It is an 18-pin device that accepts the inputs from the two sensors, and produces an eight-bit digital output. It can receive the signals from the sensors in Figure 8-19, and produce a digital output proportional to the field gradient. Also, if you want a dc output, then the same sort of D/A converter used in the magnetometer of Figure 8-17 can also be used for the gradiometer.

Figure 8-21 shows a method of using digidyning to make a very sensitive gradiometer at low cost. The outputs of the two FGM-3 sensors are fed to the D-input and clock (CLK) input of a Type-D flip-flop. The output of the Type-D flip-flop is fed to an F/V converter such as the LM-2917 device discussed earlier.

## Interfacing FGM-x Series Devices Via Microcontrollers

Microcontroller chips bring some of the advantages in single integrated circuit or small assembly form (see Chapter 18). Figure 8-22 shows a method for interfacing the FGM-x/SCL-00x series of devices to a device such as the Parallax, Inc. BASIC Stamp, or the Micromint PicStic product. If the calculations cannot be done in the microcontroller, then use the serial output capability to send the eight-bit data to a personal computer.

# Chapter 9

# Sensor Interfaces

While the number of different sensors and transducers available to the instrument designer is quite large and varied, from an interface circuit point of view there are only a few different basic forms of sensor output circuit configuration. These forms must be properly matched to the input of the circuit that follows the sensor, or trouble will result. In this chapter we will look at the basic types of sensor output circuit and their appropriate analog interface type.

## Sensor/Transducer Output Circuit Forms

Figures 9-1a through 9-1f show an array of several different forms of sensor circuit. In each circuit, a current source, source resistance (R) and (in some) a voltage source is shown. Figure 9-1a shows the standard single-ended grounded sensor.[2] The term "single-ended" means that the signal is taken between a single point and a common (or ground) point. In this case, one side of the sensor circuit is grounded. If neither side is grounded, then the sensor is said to be single-ended floating sensor (Figure 9-1b). In the single-ended sensor, the output signal is referenced either to ground or a single common, nongrounded, point. This form of circuit is sometimes subject to massive interference from external fields, especially in the presence of strong AF, RF or 60 Hz power line fields. A variant on the single-ended floating sensor is the "single-ended floating driven off ground" sensor of Figure 9-1c.

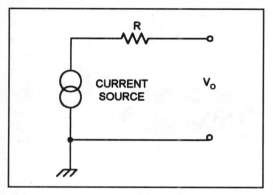

*Figure 9-1a. Single-ended unbalanced.*

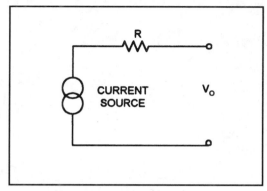

*Figure 9-1b. Singled-ended balanced.*

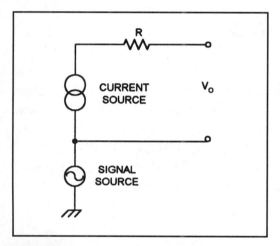

*Figure 9-1c. Signal-ended unbalanced with signal source.*

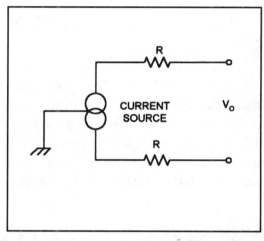

*Figure 9-1d. Split unbalanced input.*

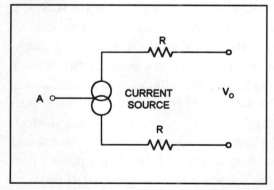

*Figure 9-1e. Split balanced input.*

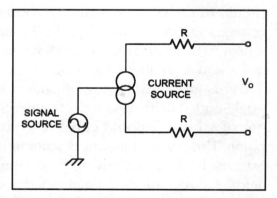

*Figure 9-1f. Split balanced with single-ended signal source.*

If a sensor drives the output through equal resistances, then it is said to be balanced. Figure 9-1d shows an example of a balanced grounded sensor. The signal is the difference between the signals between the two points, both of which are referenced to ground. In this form of output circuit, the sensor is referenced to ground through two equal resistances (both designated R). The version shown in Figure 9-1e is an example of a balanced floating sensor. That is, the sensor is connected to a non-grounded common point, and outputs through two equal resistances (R). The important point in the balanced floating sensor is that it is both balanced and ungrounded. Finally, in Figure 9-1f we see the balanced driven off ground sensor.

The different forms of sensor output must be matched to an appropriate load circuit, which almost always means an amplifier circuit. But amplifiers have different types of input configuration, and these must be accounted for in the design of the interface.

## Amplifier Input Circuits

The output circuit of the sensor is usually connected to a signal processing circuit, most frequently an amplifier of some sort (although certain other circuits are also used occassionally). Unfortunately, there are several types of amplifier input circuit, and not all sensors can be easily interfaced with all types of amplifier input circuit. Figure 9-2 shows four basic types of input circuit. In Figure 9-2a is the Type I circuit, i.e. one that is a single-ended input amplifier. The input circuit is modeled as a resistance to ground. In Figure 9-2b is the Type II circuit, which is modeled as a pair of differential inputs that each see equal resistances to ground. In both cases, the output circuit is a voltage source in series with an output resistance. Figure 9-2c shows the Type III input circuit, which is single-ended floating and shielded. The input resembles the regular single-ended input (Figure 9-2a), but the input is grounded and protected from interference by a shield. Finally, in Figure 9-2d we see the Type IV input circuit. This circuit resembles the Type II, except that the input circuit is protected by a shield, is floating, and is guarded.

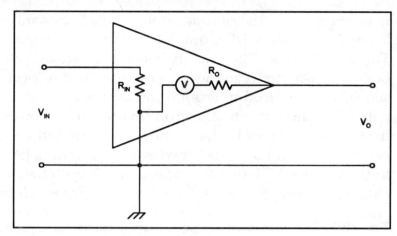

*Figure 9-2a. Type I amplifier.*

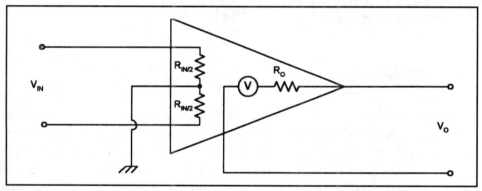

*Figure 9-2b. Type II amplifier.*

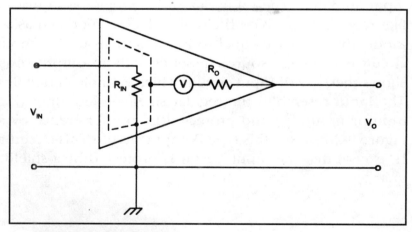

*Figure 9-2c. Type III amplifier.*

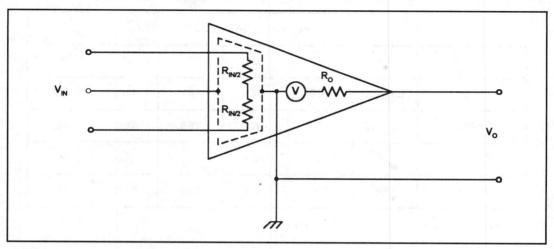

*Figure 9-2d. Type IV amplifier.*

## Matching Sensors and Amplifiers

One cannot simply connect the various forms of sensor to the various types of amplifier input circuit willy-nilly without some thought about the matter. Figure 9-3 shows a general table relating the sensor and amplifier circuits. A "YES" in a block means the combination (row vs. column) is recommended. A "NO" means that there are problems in that particular combination, so it is not recommended.

There are two combinations where it may or may not work, depending on the circumstances, so some degree of caution is required. For example, mixing a Type I input circuit with a Form A sensor output circuit requires consideration of signal levels. Do not use it when the output of the sensor is in the microvolt or millivolt range. Also, it is not a good idea to mix two grounds, i.e. one each on the amplifier and the sensor. Either eliminate one of the grounds, or join them together in a "single point" (also called "star") ground (of which, more later).

A similar problem occurs when interfacing a Form A sensor and a Type II amplifier input. Some differential amplifiers can be converted into a single-ended amplifier, but one must be certain in each case.

| INPUT CIRCUIT TYPE | SENSOR SIGNAL SOURCE FORM | | | | | |
|---|---|---|---|---|---|---|
| | A | B | C | D | E | F |
| TYPE I | SEE TEXT | YES | NO | NO | YES | NO |
| TYPE II | SEE TEXT | YES | NO | YES | YES | NO |
| TYPE III | YES | YES | YES | | YES | NO |
| TYPE IV | YES | YES | YES | YES | YES | YES |

*Figure 9-3 . Source/amplifier interface table.*

# Building the Analog Subsystem

Computer-based electronic instruments generally use some form of analog input signal that is digitized before being input to a digital computer. Although at one time quite rare (because of the cost of the computers), the advent of the desktop personal computer, single-board embedded computers and devices such as microcontroller chips has created many opportunities for instrument designers. And those opportunities were taken, for the instrumentation business is jammed with computerized devices. But most of the sensors that those instruments use are analog in nature, so are not compatible with the computer. The purpose of the analog subsystem is to acquire and process those analog sensor signals to make them fit for the computer.

## A Generic Analog Subsystem

The particulars of any specific analog subsystem will differ from others, but for purposes of discussion we will assume a generic system such as Figure 10-1. In this instrument design, a sensor is used to acquire information about some measured parameter. This signal is passed to an amplifier, may be filtered to eliminate noise and artifact, and is then treated in a post-amplifier before being applied to the input of an analog-to-digital (A/D) converter. While that seems quite simple in overall concept, "the devil", as they say, "is in the details."

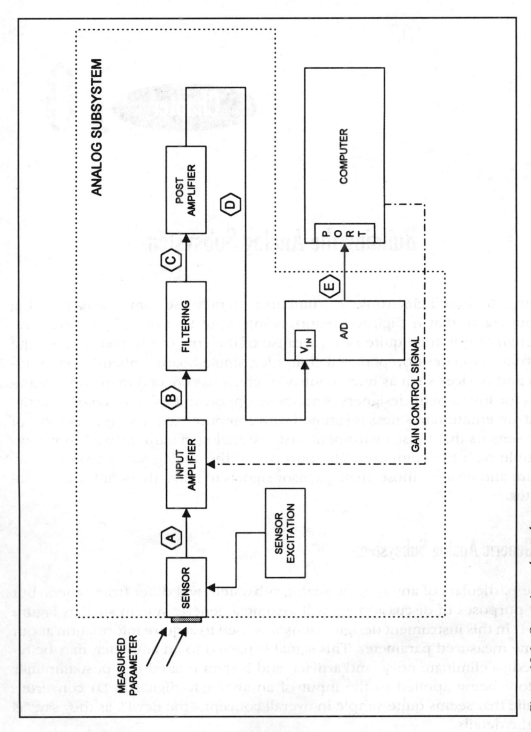

*Figure 10-1. Generic analog subsystem.*

Let's make another assumption. The sensor is a piezoresistive strain gage that measures fluid pressure. Most such sensors are based on the Wheatstone bridge circuit, and have strain gauge elements for all four arms of the bridge.

The bridge requires an excitation voltage. This voltage may be ac or dc, but for the most part dc is used. The value of this voltage must be high enough to generate a decent signal level, but low enough to prevent self-heating of the sensor elements. For most pressure sensors values of dc excitation potential is usually specified as 2.5 Vdc to 10 Vdc, with 5 Vdc being very common.

Sensor signals from bridges tend to be quite small. The sensitivity factor ($\Psi$) is usually measured in terms of the output voltage per unit of excitation voltage per unit of applied stimulus. For example, one fluid pressure sensor produces an output signal that is 30 $\mu V$ per volt excitation per Torr of pressure (note: 1 Torr = 1 mmHg), or

$$\Psi = \frac{30\ \mu V}{V - Torr} \qquad \qquad eq.\ (10\text{-}1)$$

Suppose we want to build a pressure instrument that will measure fluid pressures over the range 0 to 400 Torr, with a resolution of 0.5 Torr. In order to achieve a specified resolution, it is often necessary to build the system with a resolution that is five to ten times the target resolution. If we select the 5× rule, then we will want to be able to resolve 0.1 Torr.

One of the first questions to answer is the bit-length of the A/D converter needed for this instrument. If we want to measure 0-400 Torr in 0.1 Torr increments, then we need to be able to digitally resolve 400/0.1 states, or about 4,000 states. An 8-bit A/D converter only resolves 256 states, and the 10-bit A/D only 1,024. The 12-bit A/D converter, however, will resolve 4,096 states. This is close enough to 4,000 to permit us to not waste states above the range. If we select a 12-bit A/D converter, then each 1-LSB increment of the A/D converter represents 0.1 Torr, with a maximum range of 409.6 Torr.

The next question is to select an input range for the A/D converter. Here we must consider the excitation voltage, not just the A/D converter. The unipolar converter ranges are typically 0 to 2.5 volts, 0 to 5.0 volts and 0 to 10 volts. Let's take the 5 Vdc case first.

If we use a 12-bit, 0 to 5 Vdc, A/D converter, then each 1-LSB step represents 1/4,096 of the whole, or a proportion of 0.00024 of the whole. Since we need to account for zero, we can reduce the maximum voltage applied to the A/D. In the case of a 5 volt A/D converter, 1-LSB represents $0.00024 \times 5$ Vdc, or 0.00122 volts. If we reduce the maximum input range by 1-LSB, then the maximum A/D converter becomes $5.00 \times 0.00122$ volts, or 4.9988 volts.

If the maximum range is 409.6 Torr, and the sensitivity of the sensor is 30 µV/V-T, then the output voltage is:

$$V_O = \frac{30 \, \mu V}{V \times Torr} \times 5V \times \frac{10^{-6} \, V}{\mu V} \times 409.6 \, Torr = 0.0614 \, volts$$

eq. (10-2)

The gain required is the quotient of the maximum A/D input voltage over the maximum sensor output voltage, or

$$A_V = \frac{4.9988 \, V}{0.0614} = 81.42$$

eq. (10-3)

An overall analog subsystem gain of 81.42 will raise the maximum signal from the sensor to the maximum allowable A/D input voltage.

But now suppose the excitation voltage applied to the sensor drifts 10 µV from 5.000 Vdc to 5.010 Vdc. According the sensitivity equation, this means that the sensor output will change from 0.0614 volts at 409.6 Torr to 0.06156 volts, or a difference of 0.06156 - 0.0616 = 0.00016 volts (0.16 µV). This value is 0.16 µV/1.22 µV, or about 13 percent of the A/D converter's 1-LSB shift. This value is well within the shift that can be tolerated for a 0.1 Torr resolution.

The general rule is to use an A/D converter with a bit-length sufficient to cover the range of measurement within the range of possible drift of excitation potentials. The trade-off is whether to control the excitation potential drift or the A/D bit-length. In the case cited, the drift is 10 µV, which is well within the capabilities of most common reference sources used for excitation potentials. In fact, it would be possible to control the excitation potential to within ±1 µV without too much cost.

Another trade-off to consider is the A/D bit-length vs. amplifier gain. It is problematic to waste A/D input range. If we wish to make a multiple range

instrument, then we can digitally control the input amplifier gain to always fill up the A/D range regardless of the maximum value of the applied stimulus.

The filtering is used to reduce the bandwidth of the analog subsystem to the bandwidth of the applied signal, with the A/D sampling rate taken into consideration (the filtering can prevent aliasing). The nature of the signal determines the Fourier components, which in turn tell us the maximum bandwidth of the overall subsection. If the filter has gain, then its gain is part of the overall gain of the circuit. For example, suppose the filter has a gain of 5×. If the overall gain of the circuit is 81.42, then the gain of the amplifier and post-amplifier together must be 81.42/5, or 16.26. Post amplifiers typically have a gain of unity (1), so in that case the amplifier will have all of the gain. If the post-amplifier has gain, then the amplifier gain is 16.26 divided

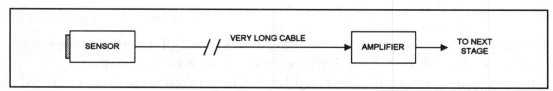

*Figure 10-2. Sensor system with a high potential for noise pick-up.*

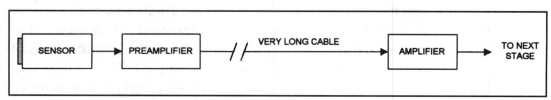

*Figure 10-3. Improved sensor system.*

by the post-amplifier gain. In some cases, the filter imposes a loss rather than a gain. In that case, the gain of the post-amplifier is set to overcome the loss of the filter.

The principal function of the post-amplifier is to provide buffering and level shifting for the output signal. In some cases, when the gain need only be one, the post-amplifier might be a unity gain noninverting follower based on the operation amplifier (see Chapter 12). In other cases, there might be some level or baseline shifting to accommodate dc Balance or Baseline controls.

The architecture of the analog subsystem sometimes makes a profound difference in the performance of the system. Consider Figure 10-2. In this case, a sensor is connected to its amplifier through a very long cable. Or alternatively, a Wheatstone bridge produces a balanced output. If the two wires that send the signal to the amplifier are long, then the possibility of common mode interference becomes more likely. A way to overcome that type of problem is to add a preamplifier at the sensor site (Figure 10-3). It will boost the signal relative to the noise picked up in the cabling, so therefore improve the signal-to-noise ratio.

Figure 10-4 shows a sensor system that I built for a research scientist. The sensor (Figure 10-4a) is a Grass FT-3. It produces a differential output of a few microvolts per gram-force (1 g-F = 980 dynes). At 1 g-F, the sensor output was around 30 μV.

The sensor was connected to the differential input of an oscilloscope and strip-chart recorder. But the balance of the pair was not all that good, so differences in the signal picked up by the two wires were translated as a relatively large common mode 60 Hz interference signal. This signal represented around 20 percent of the overall amplitude of the sensor output.

The solution to the problem was to build a gain of 100 differential amplifier and place it at the sensor end of the system. Figure 10-4b shows the amplifier. It was built in a standard instrument box, and used a connector that matched the connector on the sensor. Figure 10-4c shows the amplifier mated to the sensor. The output of the amplifier was 3 μV single-ended, which was sufficient to overcome the 60 Hz interference when shielded cable was used.

*Figure 10-4a. Sensor.*

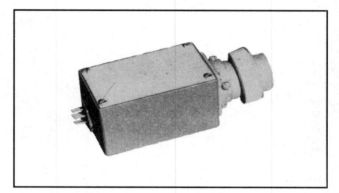

*Figure 10-4b. Amplifier.*

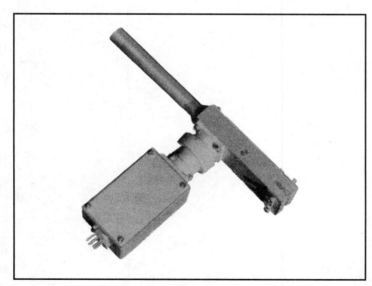

*Figure 10-4c. Mated sensor and amplifier.*

# Chapter 11

# Introduction to Analog Amplifiers

Sensor output signals are typically analog voltages or currents, and are rarely directly suitable for further use without amplification. In the next three chapters we will examine some specific amplifier circuits that are widely used with sensors, as well as in other instrumentation and control systems. The goal is to equip you with an understanding sufficient to perform some circuit design tasks. The topic of this chapter is an introduction to amplifiers, while in Chapter 12 we will take a look at the inverting and noninverting follower configurations of the operational amplifier. In Chapter 13 we will take a look at the differential and isolation amplifiers that are so widely used in sensor based technology.

The operational amplifier is, by far, the most useful and most widely applied linear integrated circuit (IC) device on the market today. It was also one of the first IC devices on the market. No supposed instrumentation or sensor circuit designer is properly educated until he or she can design with op-amps (they are that ubiquitous!). These devices, by and large, define the field of linear IC electronics. Because of their considerable importance in electronic sensors and instrumentation circuits, this chapter will discuss the IC operational amplifier in depth.

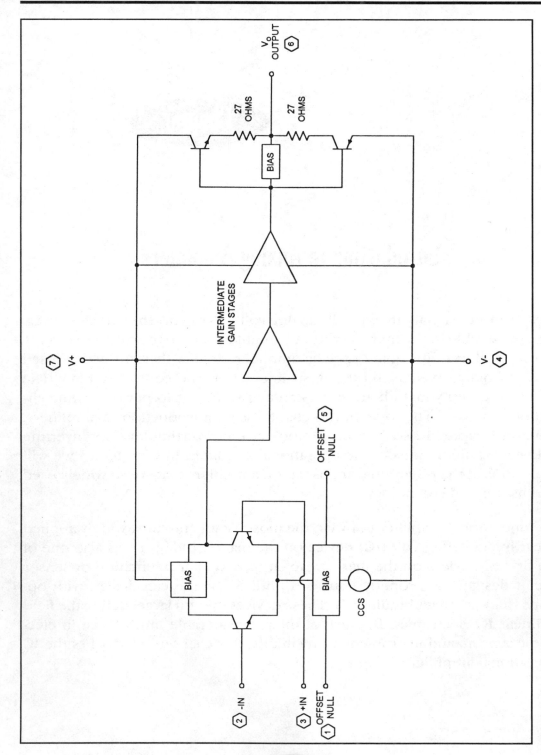

*Figure 11-1. Block diagram to operational amplifier.*

# The Operational Amplifier

The original operational amplifiers (circa 1948) were designed to perform mathematical operations in analog computers; hence the name operational amplifiers. Their use in other applications follow from the fact that the op-amp is basically a very good dc differential amplifier with extremely high gain. The immense benefits and extreme flexibility of the op-amp derive from that one simple fact. Applications for the operational amplifier are found in instrumentation, process monitoring and control, servo control systems, signals processing, communications, measuring and testing circuits, alarm systems, medicine, life and physical sciences, and even in some digital computers. It is only somewhat exaggerated to call the operational amplifier a "universal linear amplifier."

Operational amplifiers are probably the most widely used linear integrated circuits on the market today. They are, without a doubt, the most flexible linear IC devices available because one can manipulate the overall forward transfer function by manipulating the feedback network properties. Other "universal" linear IC amplifiers have failed to overcome the op-amp in either sales or usefulness.

Figure 11-1 shows a simplified schematic diagram of the internal circuitry of the popular, and very low-cost, 741 operational amplifier device. There are three main sections to this circuit: input amplifier, gain stages, and output amplifier.

The output stage is a complementary symmetry push-pull dc power amplifier. Typical devices produce from 50 to 500 milliwatts of output power, depending on design. The output stage operates as a push-pull amplifier because transistors Q9 and Q10 have opposite polarities: Q9 is NPN and Q10 is PNP. Output signal is taken from the junction of the Q9/Q10 emitters. If the two transistors conduct an equal amount of current, then the net output voltage is zero.

The intermediate gain stages are shown here in simplified block diagram form. These stages provide the high gain required, some level translation, and (in the case of 741-family devices) some internal frequency compensation.

The input stage is a dc differential amplifier made from bipolar transistors. Although there are a few single-input devices on the market sold as "operational" amplifiers, they are in reality only high gain dc amplifiers, not true "op-amps." The reason is that a true operational amplifier must be able to perform a wide range of mathematical operations, and that ability requires both inverting and noninverting input functions. This same reason is also used to explain why bipolar dc power supplies must be used in op-amp circuits. The results (i.e. output voltages) may be either zero, positive or negative, and the device must accommodate all three of these possibilities. An amplifier must have differential inputs, and bipolar output polarity, in order to operate in all four quadrants of the Cartesian coordinate system.

## DC Differential Amplifiers

Figure 11-2a shows a simplified dc differential amplifier circuit. Two transistors (Q1 and Q2) are connected at their emitters to a single constant current source (CCS), I3. Because current I3 cannot vary, changes in either I1 or I2 will also affect the other current. For example, an increase in current I1 means a necessary decrease in current I2 in order to satisfy Equation 11-1:

$$I3 = I1 + I2 \qquad\qquad eq. (11\text{-}1)$$

Where:

I3 is the constant current.

I1 is the C-E current of transistor Q1.

I2 is the C-E current of transistor Q2.

Although the collector and emitter currents for each transistor are not actually equal to each other (differing by the amount of the base current), we may conveniently neglect $I_{B1}$ and $I_{B2}$ for the sake of this discussion, and assume that $I1 = I_{C1} = I_{E1}$, and $I2 = I_{C2} = I_{E2}$ (even though they are not, in fact, equal in real circuits).

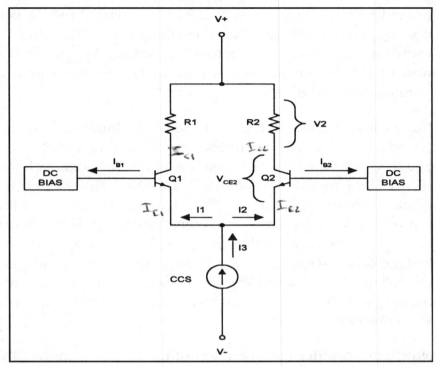

*Figure 11-2a.
Differential
input amplifier.*

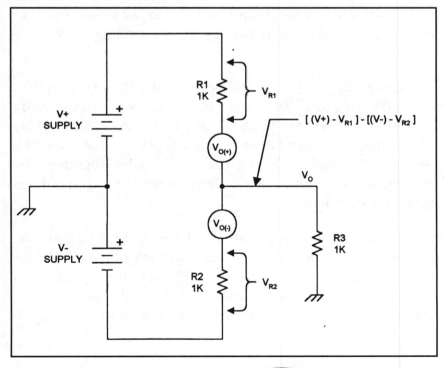

*Figure 11-2b.
Equivalent
output circuit.*

Consider two voltage drops created by current I2 (which is the C-E current flowing in transistor Q2): V2 is the voltage drop I2R2, while $V_{CE2}$ is the collector to emitter voltage drop across Q2. Voltage $V_{CE2}$ depends on the conduction of Q2, which in turn is determined by the signal applied to the noninverting input (+IN).

If the voltages applied to both -IN and +IN inputs are equal, then I1 and I2 are equal. In that case, the quiescent values of V2 and $V_{CE2}$ are approximately equal (as determined by internal bias networks) so the relative contributions of V- and V+ to output potential $V_o$ are equal. How does this work? A model circuit is shown in Figure 11-2b to demonstrate how $V_o$ is formed. The output voltage $V_o$ is the sum of two contributors: $V_{o(-)}$ is the contribution from V- and $V_{o(+)}$ is the contribution from V+. These voltages are derived from the voltage drops across R1 and R2, respectively, and in our model represents the voltage drops V2 and $V_{CE2}$ above. As long as R1 and R2 are balanced, then the sum $V_o = V_{o(-)} + V_{o(+)}$ is equal to zero. But if either R1 or R2 change, then $V_o$ will be non-zero.

Now let's consider the operation of the inverting input (-IN), which is the base terminal of transistor Q2. Recall that an inverting input produces an output signal that is 180 degrees out of phase with the input signal. In other words, as $V_{-IN}$ goes positive, $V_o$ goes negative; as $V_{-IN}$ goes negative, $V_o$ goes positive.

If a positive signal voltage is applied to the -IN input, and +IN = 0, then NPN transistor Q2 is turned on harder. The effect is an increase in current I2 because the collector-emitter resistance of Q2 drops. We now have an inequality between V2 and $V_{CE2}$: voltage V2 increases, while $V_{CE2}$ decreases. The result is that the contribution of V- to $V_o$ is greater, so $V_o$ goes negative. The base of Q2 is, therefore, the inverting input because a positive input voltage produced a negative output voltage.

If the signal voltage applied to the -IN input is negative instead of positive, then the situation changes. In that case, Q2 starts to turn off, so I2 drops. Voltage $V_{CE2}$ therefore increases and V2 decreases. The relative contribution of V+ to $V_o$ is now greater than the contribution of V-, so $V_o$ goes positive. Again inverting behavior is seen: a negative input voltage produced a positive output voltage.

Now consider the noninverting input, which is the base of transistor Q1. Recall that a noninverting input produces an output signal that is in phase with the input signal. A positive going input signal produces a positive going output signal, and a negative going input signal produces a negative going output.

Suppose -IN = 0, and a positive signal voltage is applied to +IN. In this case I1 increases. Because Equation 11-1 must be satisfied, an increase in I1 results in a decrease of I2 in order to keep I3 = I1 + I2 constant. Reducing I2 reduces V2 (which is I2R2, so the contribution of V+ to $V_o$ goes up: $V_o$ goes positive in response to a positive input voltage. This is the behavior expected of a noninverting input.

Now suppose that a negative signal voltage is applied to +IN. Now there is a decrease in the conduction of Q1, so I1 drops. Again, to satisfy Equation 11-1 current I2 increases. This condition increases V2, reducing the contribution of V+ to $V_o$, forcing $V_o$ negative. Because a negative input voltage produced a negative output voltage, we may again affirm that +IN is a noninverting input.

## Categories of Operational Amplifiers

Now that we have discussed the basic operational amplifier let's widen our discussion a bit to encompass a larger selection of devices. Some of these op-amp devices have been on the market for a long time, while others are relatively new. This discussion is intended to be representative, rather than exhaustive. Let's take a brief look at each category.

General Purpose Op-Amps. These are "garden variety" operational amplifiers that are neither special purpose nor premium devices. Most of these devices are said to be "frequency compensated," so designers tradeoff bandwidth for inherent high (but not absolute) stability. As such, the general purpose devices can be used in a very wide range of applications with very few external components. Op-amps are usually selected from this category unless some property of another class brings unique advantage to some particular application.

# Electronic Circuit Guidebook, Volume 1: Sensors

Voltage Comparators. These devices are not strictly speaking "operational amplifiers," but are based on op-amp circuitry. While all op-amps can be used as voltage comparators, the reverse is not true: IC comparators (e.g. LM-311) cannot usually be used as op-amps.

Low Input Current Op-Amps. Although ideal op-amps have zero input bias current, real devices have a small bias current due to input transistor biasing or leakage. This class of devices typically uses MOSFET, JFET or superbeta (Darlington) transistors for the input stage instead of NPN/PNP bipolar devices. Those transistors produce extremely low values of input current. The manufacturer may also elect to use a nulling method in-process that reduces input bias current. Low input current devices typically have picoampere level currents, rather than microampere or milliampere input currents found in some other devices.

Low Noise Op-Amps. These devices are optimized to reduce internally generated noise. They are generally used in the first stage or two of a cascade chain in order to improve the overall noise performance of an amplifier circuit.

Low Power Op-Amps. This category of op-amp optimizes internal circuitry to reduce power consumption. Many of these devices also operate at very low dc power supply potentials (e.g. $\pm 1.5$ Vdc).

Low Drift Op-Amps. dc amplifier circuits can experience an erroneous change of output voltage as a function of temperature. Low drift devices are internally compensated to minimize temperature drift. These devices are typically used in instrumentation circuits where drift is an important concern, especially when handling low level input signals.

Wide Bandwidth Op-Amps. Also called "video" op-amps in some literature, these devices have a very high gain-bandwidth product. One device, for example, has a G-B product of 100 MHz, compared with 300 kHz to 1.2 MHz for various 741-family devices.

Single dc Supply Op-Amps. These devices are able to provide op-amp-like behavior from a monopolar (typically V+) dc power supply. However, not all op-amp performance will be available from some of these devices because the output voltage may not be able to assume negative values.

High Voltage Op-Amps. Most op-amps operate at dc power supply potentials of ±6 to ±22 Vdc. A few devices in the high voltage category operate from ±44 Vdc supplies, and at least one proprietary hybrid model operates from ±100 Vdc power supplies.

Multiple Devices. This category requires only that more than one op-amp be included in the same package. Devices exist with two, three or four operational amplifiers in a single package. The 1458 device, for example, contains two 741-family devices in either 8-pin metal can or miniDIP packages.

Limited or Special Purpose Op-Amps. The devices in this category are designed for either a limited range of uses, or highly specialized uses. The type LM-302 buffer, for example, is an op-amp that is connected internally into the noninverting unity gain configuration. Some consumer audio devices also fit into this category. For example, there are several dual op-amp devices used for stereo audio that are optimized for audio applications.

IC Instrumentation Amplifiers. Although the instrumentation amplifier is arguably a "special purpose" device, it is sufficiently universal to warrant a class of its own. The "I.A." is a dc differential amplifier made of either two or three internal op-amps. Voltage gain can be set by either one or two external resistors.

## The "Ideal" Operational Amplifier

When you study any type of electronic device, be it transistor or integrated circuit, it is wise to start with an ideal model of that device, and then proceed to less ideal practical devices. In some cases, the practical and ideal devices are so far apart that you might wonder at the wisdom of this approach. But IC operational amplifiers, even low-cost products, so nearly approximate the ideal op-amp of textbooks that the equations actually work. The "ideal model" analysis method thus becomes extremely useful for understanding the technology, learning to design new circuits, or figuring out how an unfamiliar circuit works.

## Properties of the Ideal Operational Amplifier

The ideal op-amp is characterized by seven properties. From this short list of properties we can deduce circuit operation and design equations. Also, the list gives us a basis for examining non-ideal operational amplifiers and their defects (plus solutions to the problems caused by those defects). The basic properties of the op-amp are:

1. Infinite open-loop gain.

2. Infinite input impedance.

3. Zero output impedance.

4. Zero noise contribution.

5. Zero dc output offset.

6. Infinite bandwidth.

7. Both differential inputs stick together.

Let's take a look at these properties to determine what they mean in practical terms. You will find that some cheap op-amps only approximate some of these ideals, while for others on the list the approximation is extremely good.

Infinite Open-Loop Gain ($A_{vol}$). The open-loop gain of any amplifier is its gain without either negative or positive feedback. By definition, negative feedback is a signal fed back to the input 180 degrees out of phase. In operational amplifier terms this means feedback between the output and the inverting input.

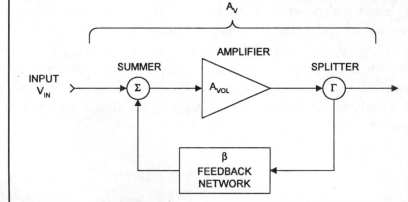

*Figure 11-3.*
*Feedback amplifier*
*block diagram.*

Negative feedback has the effect of reducing the open loop gain ($A_{vol}$) by a factor ($\beta$) that depends on the transfer function and properties of the feedback network. Figure 11-3 shows the basic configuration for any negative feedback amplifier. The transfer equation for any circuit is the ratio of the output function and the input function. The transfer function of a voltage amplifier is, therefore, $A_{vol} = V_o/V_{in}$. In Figure 11-3 the term $A_{vol}$ represents the gain of the amplifier element only, i.e. the gain with the feedback network disconnected. The overall transfer function of this circuit, i.e. with both amplifier element and feedback network ($\beta$) in the loop, is defined as:

$$A_v = \frac{A_{vol}}{1 + A_{vol}\beta} \qquad\qquad eq.\ (11\text{-}2)$$

Where:

A$_v$ is the closed-loop gain.

A$_{vol}$ is the open-loop gain.

$\beta$ is the transfer equation for the isolated feedback network.

In the ideal op-amp, $A_{vol}$ is infinite, so (as the math reveals) the voltage gain is a function only of the characteristics of the feedback network. In real op-amps, the value of the open-loop gain is not quite infinite, but it is extremely high. Typical values range from 20,000 in low-grade consumer audio models to more than 2,000,000 in premium units (in common 741 devices it is typically 200,000 to 300,000).

Infinite Input Impedance. This property implies that the op-amp input will not load the signal source. The input impedance of any amplifier is defined as the ratio of the input voltage the input current: $Z_{in} = V_{in}/I_{in}$. When the input impedance is infinite, therefore, we must assume that the input current is zero. Thus, an important implication of this property is that the operational amplifier inputs neither sink nor source current. In other words, it will neither supply current to an external circuit, nor accept current from an external circuit. We will depend upon an implication of this property ($I_{in} = 0$) to perform the simplified circuit analysis used in this chapter to determine the gain equations.

Real operational amplifiers have some finite input current other than zero. In low-grade devices this current can be substantial (e.g. >1 mA), and will cause a large output offset voltage error in medium and high gain circuits. The primary source of this current is the base bias currents from the NPN and PNP bipolar transistors used in the input circuits. Certain premium op-amps that feature bipolar inputs reduce this current to nanoamperes ($10^{-9}$ A) or picoamperes ($10^{-12}$ A). In op-amps that use field effect transistors (FET) in the input circuits, on the other hand, the input impedance is quite high due to the very low leakage currents normally found in FET devices.

The JFET input devices are typically called "BiFET" op-amps, while the MOSFET input models are called "BiMOS" devices. The CA-3140 device is a BiMOS op-amp in which the input impedance approaches 1.5 terraohms (i.e. $1.5 \times 10^{12}$ ohms) - which is near enough to "infinite", or to make the input circuits of those devices approach the ideal.

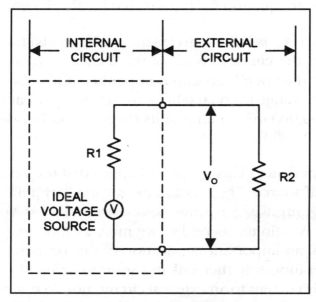

*Figure 11-4. Output circuit model.*

Zero Output Impedance. A voltage amplifier (of which class, the op-amp is a member) ideally has a zero output impedance. All real voltage amplifiers, however, have a non-zero (but low) output impedance. Figure 11-4 represents any voltage source (including an amplifier output) and its load (external circuit). Potential V is a perfect internal voltage source with no internal resistance; resistor R1 represents the internal resistance of the source, and R2 is the load. Because the internal resistance (which in amplifiers is usually called "output resistance") is in series with the load resistance, the output voltage ($V_o$) that is available to the load is reduced by the voltage drop across R1. Thus, the output voltage is given by:

$$V_o = \frac{V R2}{R1 + R2}$$

eq. (11-3)

It is clear from the above equation that the output voltage will equal the internal source voltage only when the output resistance of the amplifier (R1) is zero. In that case, $V_o$ = V(R2/R2) = V. Thus, in the ideal voltage source, the maximum possible output voltage is obtained (and the least error) because no voltage is dropped across the internal resistance of the amplifier.

Real operational amplifiers do not have a zero output impedance. The actual value is typically less than 100 ohms, with many being in the neighborhood of 30 ohms. Thus, for typical devices the operational amplifier output can be treated as if it were ideal.

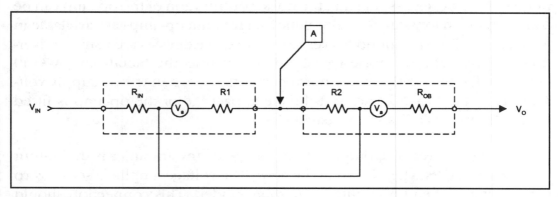

*Figure 11-5. Cascade amplifier circuit model.*

A rule of thumb used by designers is to set the input resistance of any circuit that is driven by a non-ideal voltage source output to at least ten times the previous stage output impedance. This situation is shown in Figure 11-5. Amplifier A1 is a voltage source that drives the input of amplifier A2. Resistor

R1 represents the output resistance of A1 and R2 represents the input resistance of amplifier A2. In practical terms, the circuit where R2 > 10R1 will yield results acceptably close to "ideal" for many purposes. In some cases, however, the rule R2 > 100R1 must be followed if greater precision is required.

Zero Noise Contribution. All electronic circuits, even simple resistor networks, produce noise signal at temperatures above absolute zero. A resistor creates noise due to the thermal movement of electrons in its internal resistance element material. In the ideal operational amplifier, zero noise voltage is produced internally. Thus, any noise in the output signal must have been present in the input signal as well. Except for amplification, the output noise voltage will be exactly the same as the input noise voltage. In other words, the op-amp contributed nothing extra to the output noise. This is one area where practical devices depart quite a bit from the ideal. Practical op-amps do not approximate the ideal, except for certain higher cost premium low-noise models.

Amplifiers use semiconductor devices that create not merely resistive noise (as described above), but also create other noise of their own. There are a number of internal noise sources in semiconductor devices, and any good text on transistor theory will give you more information on them. For present purposes, however, assume that the noise contribution of the op-amp can be considerable in low signal level situations. Premium op-amps are available in which the noise contribution is very low, and these devices are usually advertised as premium low-noise types. Others, such as the metal-can CA-3140 device, will offer relatively low noise performance when the dc supply voltages are limited to ±5 Vdc, and the metal package of the op-amp is fitted with a flexible "TO-5" style heatsink.

Zero Output Offset. The output offset voltage of any amplifier is the output voltage that exists when it should be zero. The voltage amplifier sees a zero input voltage when the inputs are both grounded. This connection should produce a zero output voltage. If the output voltage is non-zero, then there is an output offset voltage present. In the ideal op-amp, this offset voltage is zero volts - but real op-amps exhibit at least some amount of output offset voltage. In the real IC operational amplifier the output offset voltage is non-zero, although it can be quite low in some cases.

Infinite Bandwidth. The "ideal" op-amp will amplify all signals from dc to the highest ac frequencies. In real op-amps, however, the bandwidth is sharply limited. There is a specification called the gain-bandwidth (G-B) product, which is symbolized by $F_t$. This specification is the frequency at which the voltage gain drops to unity (1). The maximum available gain at any frequency is found from dividing the maximum required frequency into the gain-bandwidth (G-B) product. If the value of Ft is not sufficiently high, then the circuit will not behave in classical op-amp fashion at some higher frequencies within the range of interest.

Some op-amps have G-B products in the 10 to 20 MHz range. Others, on the other hand, are quite limited. A few special devices have G-B products up to 600 MHz. The 741-family of devices is very limited, such that the device will perform as an op-amp only to frequencies of a few kilohertz. Above that range, the gain drops off considerably. But in return for this apparent limitation, we obtain nearly unconditional stability; such op-amps are said to be "frequency compensated." It is the frequency compensation of those devices both reduces the G-B product and provides the inherent stability. Non-compensated op-amps will yield wider frequency response, but only at the expense of a tendency to oscillate. Those op-amps may spontaneously oscillate without any special encouragement if certain precautions are not taken in the circuit design.

Differential Inputs Stick Together. Most operational amplifiers have two inputs: an inverting (-IN) input and an noninverting (+IN) input. "Sticking together" means that a voltage applied to one of these inputs also appears at the other input. This voltage is real - it is not merely some theoretical device used to evaluate circuits. If you apply a voltage to, say, the inverting input, and then connect a voltmeter between the noninverting input and the power supply common, then the voltmeter will read the same potential on the noninverting as it did on the inverting input. The implication of this property is that both inputs must be treated the same mathematically. This fact will make itself felt when we discuss the concept of "virtual" as opposed to actual grounds, and again when we deal with the noninverting follower circuit configuration.

The inverting follower produces an output signal that is 180 degrees out of phase with its input signal. The noninverting follower, as you might expect, produces an output signal that is in phase with its input signal. Almost all other operational amplifier circuits are variations on either inverting or noninverting follower circuits. Understanding these two configurations will allow you to understand, and either design or modify, a wide variety of different circuits using IC operational amplifiers.

## Standard Operational Amplifier Parameters

Understanding operational amplifier circuits requires knowledge of the various parameters given in the specification sheets. The list below represents the most commonly needed parameters.

Open-Loop Voltage Gain ($A_{vol}$). Voltage gain is defined as the ratio of output voltage to input signal voltages ($V_o/V_{in}$), which is a dimensionless quantity. The open-loop voltage gain is the gain of the circuit without feedback (i.e. with the feedback loop open). In an ideal operational amplifier $A_{vol}$ is infinite, but in practical devices it will range from about 20,000 for low-cost devices to over 2,000,000 in premium devices.

Large Signal Voltage Gain. This gain figure is defined as the ratio of the maximum allowable output voltage swing (usually one to several volts less than V- and V+) to the input signal required to produce a swing of ±10 volts (or some other standard).

Slew Rate. This parameter specifies the ability of the amplifier to transition from one output voltage extreme to the other extreme, while delivering full rated output current to the external load. The slew rate is measured in terms of voltage change per unit of time. The 741 operational amplifier, for example, is rated for a slew rate of 0.5 volts per microsecond (0.5 V/µS). Slew rate is usually measured in, and specified for, the unity gain noninverting follower configuration.

Common Mode Rejection Ratio (CMRR). A common mode voltage is one that is presented simultaneously to both inverting and noninverting inputs. In an ideal operational amplifier, the output signal resulting from the common mode voltage is zero, but in real devices it is non-zero. The CMRR is the measure of the device's ability to reject common mode signals, and is expressed as the ratio of the differential gain to the common mode gain. The CMRR is usually expressed in decibels, with common devices having ratings between 60 dB and 120 dB (the higher the number, the better the device).

Power Supply Rejection Ratio (PSRR). Also called power supply sensitivity, the PSRR is a measure of the operational amplifier's insensitivity to changes in the power supply potentials. The PSRR is defined as the change of the input offset voltage (see below) for a one volt change in one power supply potential (while the other is held constant). Typical values are in microvolts or millivolts per volt of power supply potential change.

Input Offset Voltage. The voltage required at the input to force the output voltage to zero when the input signal voltage is zero. The output voltage of an ideal operational amplifier is zero when $V_{in}$ is zero.

Input Bias Current. This current is the current flowing into or out of the operational amplifier inputs. In some sources, this current is defined as the average difference between currents flowing in the inverting and noninverting inputs.

Input Offset (Bias) Current. The difference between inverting and noninverting input bias current when the output voltage is held at zero.

Input Signal Voltage Range. The range of permissible input voltages as measured in the common mode configuration.

Input Impedance. The resistance between the inverting and noninverting inputs. This value is typically very high: 1 megohm in low-cost bipolar operational amplifiers and over $10^{12}$ ohms in premium BiMOS devices.

Output Impedance. This parameter refers to the "resistance looking back" into the amplifier's output terminal, and is usually modeled as a resistance between output signal source and output terminal. Typically the output impedance is considerably less than 100 ohms.

Output Short Circuit Current. The current that will flow in the output terminal when the output load resistance external to the amplifier is zero ohms (i.e. a short to common).

Channel Separation. This parameter is used on multiple operational amplifier integrated circuits, i.e. devices in which two or more operational amplifiers sharing the same package with common power supply terminals.

The separation specification tells us something of the isolation between the op-amps inside the same package, and is measured in decibels (dB). The 747 dual operational amplifier, for example, offers 120 dB of channel separation. From this specification we may imply that a 1 microvolt change will occur in the output of one of the amplifiers when the other amplifier output changes by 1 volt (20 LOG[1 V/1 $\mu$V] = 120 dB).

## Minimum and Maximum Parameter Ratings

Operational amplifiers, like all electronic components, are subject to certain maximum ratings. If these ratings are exceeded, then the user can expect either premature - often immediate - failure, or at least unpredictable operation. The ratings mentioned below are the most commonly used.

Maximum Supply Voltage. This potential is the maximum that can be applied to the operational amplifier without damaging the device. The operational amplifier uses V+ and V- dc power supplies that are typically $\pm$18 Vdc, although some exist with much higher maximum potentials.

Power Dissipation ($P_d$). This rating is the maximum power dissipation of the operational amplifier in the normal ambient temperature range (80° C in commercial devices, and 125° C in military-grade devices). A typical rating for op-amps is 500 milliwatts (0.5 watts).

Maximum Power Consumption. The maximum power dissipation, usually under output short circuit conditions, that the device will survive. This rating includes both internal power dissipation and device output power requirements.

Maximum Input Voltage. This potential is the maximum that can be applied simultaneously to both inputs. Thus, it is also the maximum common mode voltage. In most bipolar operational amplifiers the maximum input voltage is very nearly equal to the power supply voltage. There is also a maximum input voltage that can be applied to either input when the other input is grounded.

**Differential Input Voltage.** This input voltage rating is the maximum differential mode voltage that can be applied across the inverting (-IN) and noninverting (+IN) inputs.

**Maximum Operating Temperature.** The maximum temperature is the highest ambient temperature at which the device will operate according to specifications with a specified level of reliability. The usual rating for commercial devices is 70° or 80° C, while military components must operate to 125° C.

**Minimum Operating Temperature.** There is a minimum operating temperature, i.e. the lowest temperature at which the device operates within specifications. Commercial devices operate down to either 0 or -10° C, while military components operate down to -55° C.

**Output Short-Circuit Duration.** This rating is the length of time the operational amplifier will safely sustain a short circuit of the output terminal. Many modern operational amplifiers are rated for indefinite output short circuit duration.

**Maximum Output Voltage.** The maximum output potential of the operational amplifier is related to the dc power supply voltages. Operational amplifiers have one or more bipolar PN junctions between the output terminal and either V- or V+ terminals. The voltage drop across these junctions reduces the maximum achievable output voltage. For example, if there are three PN junctions between the output and power supply terminals, then the maximum output voltage is $[(V+) - (3 \times 0.7)]$, or $[(V+) - 2.1]$ volts. If the maximum V+ voltage permitted is 15 volts, then the maximum allowable output voltage is $[(15\ V) - (2.1\ V)]$, or 12.9 volts. It is not always true, especially in older devices, that the maximum negative output voltage is equal to the maximum positive output voltage. A related rating is the maximum output voltage swing, which is the absolute value of the voltage swing from maximum negative to maximum positive.

## Practical Operational Amplifiers

Now that we have examined the ideal operational amplifier and some typical device specifications, let's turn our attention to practical devices. Because of its popularity and low cost we will concentrate on the 741 device. The 741 family also includes the 747 and 1458 "dual 741" devices. Although there are many better operational amplifiers on the market, the 741 and the members of its close family are considered the industry standard generic op-amp devices.

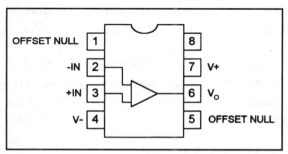

Figure 11-6a.
LM-741, 8-pin miniDIP.

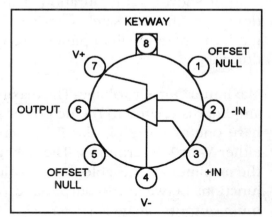

Figure 11-6b.
LM-741, 8-pin metal can.

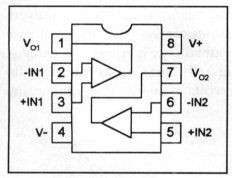

Figure 11-6c.
LM-1458 dual op-amp.

Figure 11-6 shows the two most popular packages used for the 741. Figure 11-6a is the 8-pin miniDIP package, while Figure 11-6b is the 8-pin metal can package. The 741 is also available in flatpacks and 14-pin DIP packages, although these are becoming rare today. The miniDIP pin-outs for a 1458 dual op-amp are shown in Figure 11-6c. The 741 has the following pins:

Inverting Input (-IN), Pin No. 2. The output signals produced from this input are 180 degrees out of phase with the input signal applied to -IN.

Noninverting Input (+IN), Pin No. 3. Output signals are in phase with signals applied to the +IN input terminal.

Output, Pin No. 6. On most op-amps, the 741 included, the output is single-ended. This term means that output signals are taken between this terminal and the power supply common (see Figure 11-7). The output of the 741 is said to be short-circuit proof because it can be shorted to common indefinitely without damage to the IC

V+ dc Power Supply, Pin No. 7. The positive dc power supply terminal.

V- dc Power Supply, Pin No. 4. The negative dc power supply terminal.

Offset Null, Pins 1 and 5. These two terminals are used to accommodate external circuitry that compensates for offset error voltages.

The pin-out scheme shown in Figure 11-6a is considered the de-facto "industry standard" for generic single operational amplifiers. Although there are numerous examples of amplifiers using different pin-outs other than Figure 11-6a, a very large percentage of the available devices use this scheme.

## Standard Circuit Configurations

The standard circuit configuration for 741-family of operational amplifiers is shown in Figure 11-7. The pin-outs are industry standard. The output signal voltage is impressed across load resistor RL connected between the output terminal (pin no. 6) and the power supply common. Most manufacturers recommend a 2 kohm minimum value for $R_L$. Also, note that some opera-

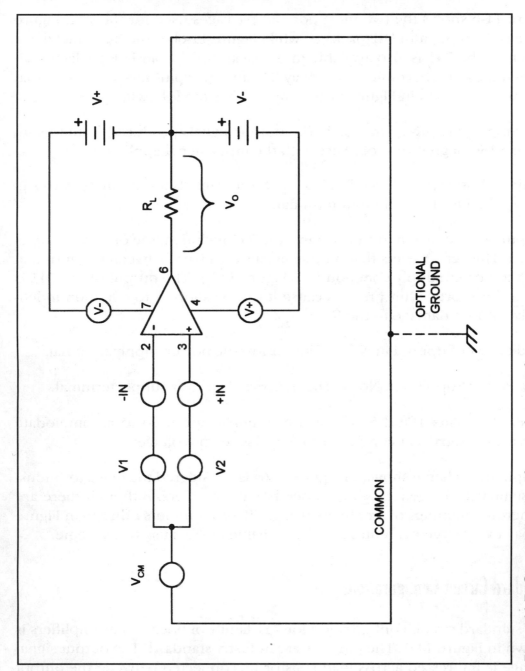

*Figure 11-7. Op-amp circuit with equivalent signals.*

tional amplifier parameters are based on a 10 kohm load resistance, so may differ for other load resistance values. Because it is referenced to common, the output is single-ended.

The ground symbol shown in Figure 11-7 indicates that it is optional. The point of reference for all measurements is the common connection between the two dc power supplies (V- and V+). Whether or not this point is physically connected to a ground, the equipment chassis or a dedicated system ground bus is purely optional in some cases, and required in others. Whether it is required or not, however, the determination is made on the basis of circuit factors other than the basic nature of the op-amp.

The V- and V+ dc power supplies are independent of each other. Do not make the mistake of assuming that these terminals are merely different ends of the same dc power supply. In fact, V- is negative with respect to common, while V+ is positive with respect to common.

The two input signals in Figure 11-7 are labelled V1 and V2. Signal voltage V1 is the single-ended potential between the common and the inverting input (-IN), while V2 is the single-ended potential between common and the noninverting input (+IN). The 741 operational amplifier is differential, as indicated by the fact that both -IN and +IN are present. Any differential amplifier produces an output that is proportional to the difference between the two input potentials. In Figure 11-7 the differential input potential $V_d$ is the difference between V1 and V2:

$$V_d = V2 - V1 \qquad\qquad eq. (11\text{-}4)$$

Signal voltage $V_{cm}$ in Figure 11-7 is the common mode signal, that is, a potential that is common to both -IN and +IN inputs. This potential is equivalent to the situation V1 = V2. In an ideal operational amplifier there will be no output response at all to a common mode signal. In real devices, however, there is some small response to $V_{cm}$. The freedom from such responses is called the common mode rejection ratio (CMRR).

# Chapter 12

# Inverting and Noninverting Amplifiers

In Chapter 11 the basic operational amplifier was introduced. Now let's take a look at the basic circuit configurations used with the op-amp. In Chapter 13 we will take a look at the differential amplifier configurations.

There are two basic configurations for operational amplifier voltage amplifier circuits: inverting and noninverting. For some reason lost in darkest antiquity, these circuits are usually called "followers." In this chapter we will examine the basic inverting follower and noninverting follower circuits.

## Inverting Follower Circuits

The inverting follower is an operational amplifier circuit configuration in which the output signal is 180 degrees out of phase with the input signal. Figure 12-1a shows the relationship between input and output signals for an inverting follower with a gain of -1. Note the phase reversal present in the output signal with respect to the input signal. In order to achieve this inversion the inverting input (-IN) of the operational amplifier is active, and the noninverting input (+IN) is grounded.

Figure 12-1b shows the basic configuration for the inverting follower (also called inverting amplifier) circuits. The noninverting input is not used, so is set to ground potential. There are two resistors in this circuit: resistor Rf is

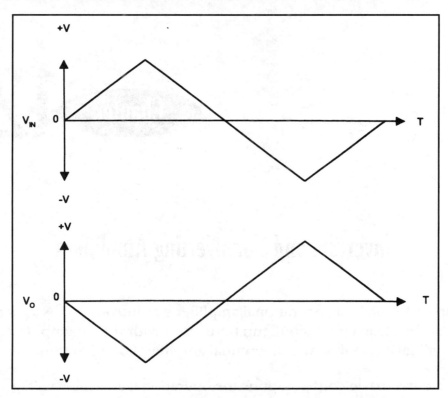

*12-1a. Input and output signals of inverting follower*

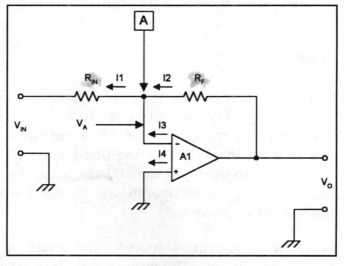

*12-1b. Inverting follower circuit.*

the negative feedback path from the output to the inverting input, while Rin is the input resistor. We will examine the $R_f/R_{in}$ relationship in order to determine how gain is fixed in this type of circuit. But first, let's take a look at the implications of grounding the noninverting input in this type of circuit.

## What's a "Virtual" Ground?

A virtual ground is a connection or circuit point that acts like a ground, even though it is not physically connected to either a truly grounded point or the circuit common point. While this definition sounds strange at first, it's not an unreasonable description of a "virtual ground." Unfortunately, that terminology is confusing and therefore leads to an erroneous implication that the virtual ground somehow doesn't really function as a ground. Let's examine the concept of a "virtual ground." Earlier you learned the properties of the "Ideal Operational Amplifier." One of those properties tells us that differential inputs "stick together." Put another way, this property means that a voltage applied to one input also appears on the other input.

In the arithmetic of op-amps, therefore, we must treat both inputs as if they are both at the same potential. This fact is not merely a theoretical device, either, for if you actually apply a potential, say 1 Vdc, to the noninverting input the same 1 Vdc potential can also be measured at the inverting input.

In Figure 12-1b, the noninverting input is grounded, so it is at zero volts potential. This fact, by the properties of the ideal op-amp, means that the inverting input of the op-amp is also at the same 0 Vdc ground potential. Since the inverting input is at ground potential, but has no physical ground connection, it is said to be at "virtual" (as opposed to "physical") ground. A virtual ground is, therefore, a point that is fixed at ground potential (0 Vdc), even though it is not physically connected to the actual ground or common of the circuit. The choice of the term "virtual ground" is unfortunate, for the concept is actually quite simple even though the terminology makes it sound a lot more abstract than it really is.

## Developing the Transfer Equation for the Inverting Follower Circuit

The transfer equation of any circuit is the output function divided by the input function. For an operational amplifier used as a voltage amplifier, therefore, the transfer function describes the voltage gain:

$$A_v = \frac{V_o}{V_{in}} \qquad\qquad eq.\ (12\text{-}1)$$

Where:

A$_v$ is the voltage gain (dimensionless).

V$_o$ is the output signal potential.

V$_{in}$ is the input signal potential (V$_o$ and V$_{in}$ are in the same units).

In the inverting follower circuit (Figure 12-1b) the gain is set by the ratio of two resistors, R$_f$ and R$_{in}$. Let's make a step-by-step analysis to see if we can find this relationship. Consider the currents flowing in Figure 12-1b. The input bias currents, I3 and I4, are assumed to be zero for purposes of analysis. This is a reasonable assumption because our model is an ideal operational amplifier. In real op-amps these currents are non-zero, and have to be accounted for, but in the analysis case we use the ideal model. Thus, in the analysis below we can ignore bias currents (assume that I3 = I4 = 0).

Remember that the summing junction (point "A") is a virtual ground, and is at ground potential because the noninverting input is grounded. Current I1 is a function of the applied input voltage, V$_{in}$, and the input resistance R$_{in}$. By Ohm's law, then, the value of I1 is:

$$I1 = \frac{V_{in}}{R_{in}} \qquad\qquad eq.\ (12\text{-}2)$$

Further, we know that current I2 is also related by Ohm's law to the output voltage, V$_o$, and the feedback resistor R$_f$ (again, because the summing junction is at 0 Vdc):

$$I2 = \frac{V_o}{R_f} \qquad\qquad eq.\ (12\text{-}3)$$

How are I1 and I2 related? These two currents are the only currents entering or leaving the summing junction (recall that I3 = 0), so by Kirchhoff's Current Law (KCL) we know that:

$$I1 + I2 = 0 \qquad \text{eq. (12-4)}$$

so,

$$I2 = -I1 \qquad \text{eq. (12-5)}$$

We can arrive at the transfer function by substituting Equations 12-3 and 12-4 into Equation 12-5:

$$I2 = -I1 \qquad \text{eq. (12-6)}$$

$$\frac{V_o}{R_f} = \frac{-V_o}{R_{in}} \qquad \text{eq. (12-7)}$$

Algebraically rearranging Equation 12-7 yields the transfer equation in standard format:

$$\frac{V_o}{V_{in}} = \frac{-R_f}{R_{in}} \qquad \text{eq. (12-8)}$$

According to Equation 12-1, the gain ($A_v$) of a circuit is $V_o/V_{in}$, so we may also write Equation 12-8 in the form:

$$A_v = \frac{-R_f}{R_{in}} \qquad \text{eq. (12-9)}$$

We have shown above that the voltage gain of an op-amp inverting follower is merely the ratio of the feedback resistance to the input resistance (-$R_f/R_{in}$). The minus sign indicates that a 180 degree phase reversal takes place. Thus, a negative input voltage produces a positive output voltage, and vice versa.

We often see the transfer equation (Equation 12-9) written to express output voltage in terms of gain and input signal voltage. The two expressions are:

$$V_o = -A_v V_{in} \qquad \text{eq. (12-10)}$$

and,

$$V_o = -V_{in} \left( \frac{R_f}{R1} \right) \qquad \text{eq. (12-11)}$$

The transfer function ($A_v = V_o/V_{in}$) can be plotted on graph paper in terms of input and output voltage. Figure 12-1c shows the plot $V_o$ vs. $V_{in}$ for an inverting amplifier with a gain of -2. In the case of a perfect amplifier the Y-intercept is 0 volts. Given the nature of Figure. 12-1c, the basic form for our purposes becomes $V_o = A_v V_{in} \pm V_{offset}$.

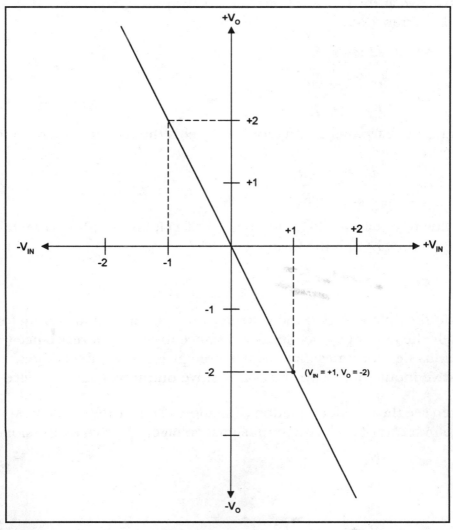

*Figure 12-1c. Transfer function.*

## Inverting Amplifier Transfer Equation by Feedback Analysis

In the section above, we developed the inverting amplifier transfer equation from the ideal model of the operational amplifier. Now let's consider the same matter from the point of view of the generic feedback amplifier to see if these equations are valid. When used in a closed-loop circuit the operational amplifier is a feedback amplifier, so feedback circuit analysis will result in the same transfer equation as the "ideal model" analysis. Figure 12-2 shows an operational amplifier with its feedback network. The overall gain of this type of amplifier is defined by the following expression:

$$A_v = \frac{A_{vol}\ C}{1 + A_{vol}\beta} \qquad\qquad eq.\ (12\text{-}12)$$

Where:

A$_v$ is the closed-loop voltage gain of the entire circuit.

A$_{vol}$ is the open-loop voltage gain of the op-amp without feedback.

C is the transfer equation of the input network.

β is the transfer equation of the feedback network.

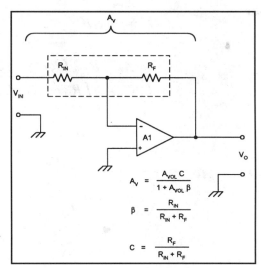

*Figure 12-2.*
*Inverting follower gain is*
*set by feedback network.*

$$A_v = \frac{A_{VOL}\ C}{1 + A_{VOL}\ \beta}$$

$$\beta = \frac{R_{IN}}{R_{IN} + R_F}$$

$$C = \frac{R_F}{R_{IN} + R_F}$$

There are two networks that must be considered in this analysis: the input network (C) and the feedback network ($\beta$); both networks are resistor voltage divider attenuators consisting of Rin and Rf, so we can expect $\beta$ and C to be fractions. The expression for the input network in Figure 12-2 is:

$$C = \frac{R_f}{R_f + R_{in}}$$  eq. (12-13)

The "C" term is needed because the input signal is attenuated by the $R_{in}/R_f$ voltage divider network. If the signal is applied directly to the inverting input, as it might be in certain other feedback amplifiers, then this input attenuation term is unity, so it disappears from Equation 12-12.

The feedback transfer equation is defined by the feedback voltage divider, $R_f/R_{in}$:

$$\beta = \frac{R_{in}}{R_f + R_{in}}$$  eq. (12-14)

We can now substitute the expressions for $\beta$ (Equation 12-14) and C (Equation 12-13) into the equation for the standard feedback amplifier (Equation 12-12):

$$A_v = \frac{A_{vol} C}{1 + A_{vol}\beta}$$  eq. (12-15)

$$A_v = \frac{A_{vol}[R_f/(R_f + R_{in})]}{1 + A_{vol}[R_{in}/(R_f + R_{in})]}$$  eq. (12-16)

$$A_v = \frac{\left(\dfrac{R_f}{R_f + R_{in}}\right)}{\left(\dfrac{1}{A_{vol}}\right) + \left(\dfrac{R_{in}}{R_f + R_{in}}\right)}$$  eq. (12-17)

Because $A_{vol}$ is infinite in ideal devices (and very high in practical devices), the term $1/A_{vol} \rightarrow 0$, so we may write Equation 12-17 in the form:

$$A_v = \frac{\left(\dfrac{R_F}{R_f + R_{in}}\right)}{\left(\dfrac{R_{in}}{R_f + R_{in}}\right)} \qquad \text{eq. (12-18)}$$

Earlier we discovered that $A_v = R_f/R_{in}$. If the feedback analysis is correct, then Equation 12-18 will be equal to $R_f/R_{in}$: Solving this relationship we invert and multiply:

$$\left(\frac{R_f + R_{in}}{R_{in}}\right) + \left(\frac{R_f}{R_f + R_{in}}\right) = \left(\frac{R_f}{R_{in}}\right) \qquad \text{eq. (12-19)}$$

$$\frac{R_f}{R_{in}} = \frac{R_f}{R_{in}} \qquad \text{eq. (12-20)}$$

Equation 12-20 demonstrates the equality of the two methods, proving that the transfer equation (Equation 12-9) derived earlier is valid.

The following equations apply to inverting followers:

$$A_v = \frac{-R_f}{R_{in}} \qquad \text{eq. (12-21)}$$

$$V_o = -A_v V_{in} \qquad \text{eq. (12-22)}$$

$$V_o = -V_{in}\frac{R_f}{R_{in}} \qquad \text{eq. (12-23)}$$

## Multiple Input Inverting Followers

We can accommodate multiple signal inputs on an inverting follower by using a circuit such as Figure 12-3. There are a number of applications of such circuits: summers, audio mixers, instrumentation and so forth. The multiple input inverter of Figure 12-3 can be evaluated exactly like Figure 12-1b, except that we have to account for more than one input circuit. Again appealing to KCL, we know that:

- $$I1 + I2 + I3 + ... + I_n = I_f \qquad\qquad eq. (12\text{-}24)$$

Also by Ohm's law, considering that summing junction "A" is virtually grounded, we know that:

- $$I1 = V1/R1 \qquad\qquad eq. (12\text{-}25)$$
- $$I2 = V2/R2 \qquad\qquad eq. (12\text{-}26)$$
- $$I3 = V3/R3 \qquad\qquad eq. (12\text{-}27)$$
- $$I_n = V_n/R_n \qquad\qquad eq. (12\text{-}28)$$
- $$I_f = V_o/R_f \qquad\qquad eq. (12\text{-}29)$$

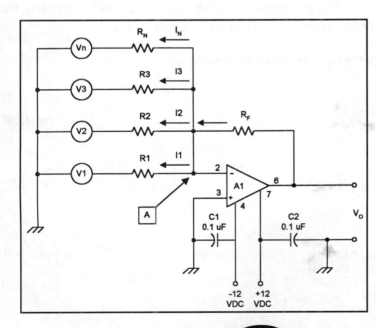

*Figure 12-3.*
*Multi-input*
*inverting follower.*

Substituting Equations 12-25 through 12-29 into Equation 12-24:

$$\left(\frac{V1}{R1}\right) + \left(\frac{V2}{R2}\right) + \left(\frac{V3}{R3}\right) + \ldots + \left(\frac{V_n}{R_n}\right) = \frac{V_o}{R_f} \qquad \text{eq. (12-30)}$$

or, algebraically rearranging Equation 12-30 to solve for $V_o$:

$$V_o = R_f\left(\frac{V1}{R1} + \frac{V2}{R2} + \frac{V3}{R3} + \ldots + \frac{V_n}{R_n}\right) \qquad \text{eq. (12-31)}$$

Equation 12-31 is the transfer equation for the multiple input inverting follower.

## Response to ac Signals

Thus far our discussion of inverting amplifiers has assumed a dc input signal voltage. The behavior of the circuit in response to ac signals (e.g. sine waves, square waves and triangle waves) is similar. Recall the rules for the inverter: positive input signals produce negative output signals, and negative input signals produce positive output signals. These relationships mean that a 180 degree phase shift occurs between input and output. The relationship is shown in Figure 12-4a.

Although the dc-coupled op-amp will respond to ac signals, there is a limit that must be recognized. If the peak value of the input signal gets too great, then output clipping (Figure 12-4b) will occur. The peak output voltage will be:

$$V_{o(peak)} = -A_v V_{in(peak)} \qquad \text{eq. (12-32)}$$

Where:

$V_o$(peak) is the peak output voltage.

$V_{in}$(peak) is the peak input voltage.

$A_v$ is the voltage gain.

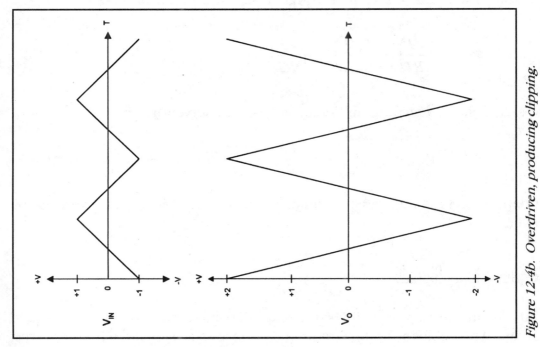

*Figure 12-4b. Overdriven, producing clipping.*

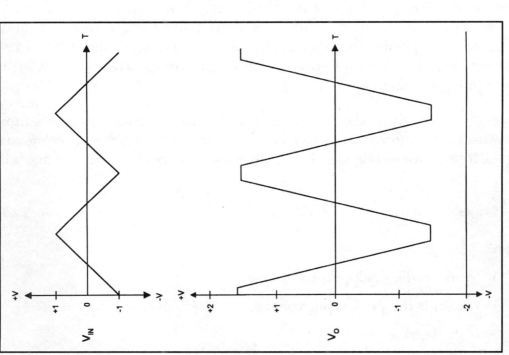

*Figure 12-4a. Input and output signals of gain-of-2 inverting amplifier.*

For every value of V- and V+ power supply potentials there is a maximum attainable output voltage, $V_{o(max)}$. As long as the peak voltage is less than this maximum allowable output potential, then the input waveform will be faithfully reproduced in the output (except amplified and inverted). But if the value of $V_{o(peak)}$ determined by Equation 12-32 is greater than $V_{o(max)}$, then clipping will occur.

In a linear voltage amplifier clipping is undesirable. The maximum output voltage can be used to calculate the maximum input signal voltage:

$$V_{in(max)} = \frac{V_{o(max)}}{A_v} \qquad \text{eq. (12-33)}$$

There are occasions when clipping is desired. For example, clipping is desired is in generating square waves from sine waves. The goal in that case is to drive the input so hard that sharp clipping occurs. Although there are better ways to realize this goal, the "overdriven clipper" square wave generator does work.

## Response to ac Input Signals with dc Offset

The case considered in the previous section assumed a waveform that is symmetrical about the zero volts baseline. In this section we will examine the case where an ac waveform is superimposed on a dc voltage. Figure 12-5 shows an inverting amplifier circuit with an ac signal source in series with a dc source. In Figure 12-6a there is a 4 volt peak-to-peak square wave superimposed on a 1 Vdc fixed potential. Thus, the non-symmetrical signal will swing between +3 volts and -1 volts.

The output waveform is shown in Figure 12-6b. With the 180 degree phase inversion and the gain of -2 depicted in Figure 12-5, the waveform will be a non-symmetrical oscillation between -6 volts and +2 volts. Because of gain $(A_v = -2)$ the degree of asymmetry has also doubled to 2 Vdc.

Dealing with ac signals that have a dc component can lead to problems at high gain and/or high input signal levels. As was true in the case of the high amplitude symmetrical signal the output may saturate at either V- or V+ power supply rails. If this limit is reached, then clipping will result. The dc compo-

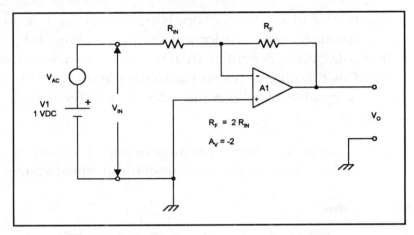

*Figure 12-5. Inverting follower with dc component in input signal.*

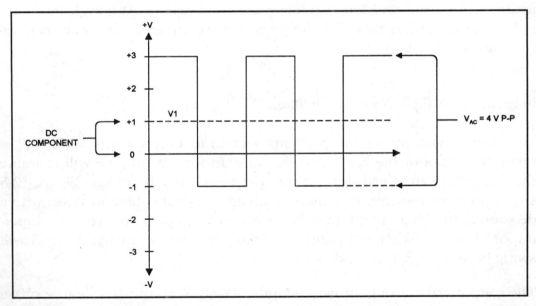

*Figure 12-6a. Input signal for Figure 12-5.*

nent is seen as a valid input signal so will drive the output to one power supply limit or the other. For example, if the op-amp in Figure 12-5 has a $V_{o(max)}$ value of $\pm 10$ volts, then (with $A_v = -2$), a $+4$ volt positive input signal will saturate the output, while the negative excursion can reach -7 volts before causing output saturation.

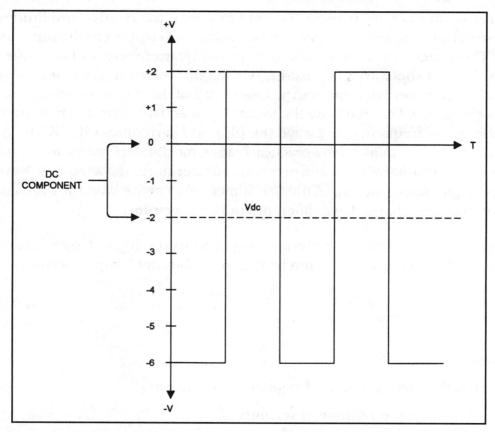

*Figure 12-6b. Output signal for Figure 12-5.*

## Response to Square Waves and Pulses

Most amplifiers respond in a congenial manner to sinusoidal and triangle waveforms. Some amplifiers, however, will exhibit problems dealing with fast risetime waveforms such as square waves and pulses. The source of these problems is the high frequency content of these waveforms.

All continuous mathematical functions (including electronic waveforms) are made of a series of harmonically related sine and cosine constituent waves (and possibly also a dc component). The sine wave consists of a single frequency, or fundamental sinusoidal wave. All non-sinusoidal waveforms, however, are made up of a fundamental sine wave plus its harmonics. The actual waveshape is determined by the number of harmonics present, which particular harmonics are present (i.e. odd or even), the relative amplitudes of those harmonics, and their phase relationship with respect to the fundamental. These factors can be deduced from the quarter-wave and/or half-wave symmetry of the wave. The listing of the constituent frequencies forms a Fourier series, and determines the bandwidth of the system required to process the signal. For example, the symmetrical square wave is made up of a fundamental frequency sine wave (F), plus odd harmonics (3F, 5F, 7F ...) up to (theoretically) infinity (as a practical matter, most square waves are "square" if the first 100 harmonics are present). Furthermore, if the square wave is truly symmetrical, then all of the harmonics are in phase with the fundamental. Other waveshapes have different Fourier spectrums.

In general, the risetime of a pulse is related to the highest significant frequency in the Fourier spectrum by the "rule-of-thumb" approximation:

$$F = \frac{0.35}{T_r} \qquad\qquad eq. (12\text{-}34)$$

Where:

F is the highest Fourier frequency in Hertz (Hz).

$T_r$ is the pulse risetime in seconds (s).

Because pulse shape is a function of the Fourier spectrum for that wave, the frequency response characteristic of the amplifier has an effect on the waveshape of the reproduced signal. Figure 12-7 shows an input pulse signal (Figure 12-7a) and two possible responses. The response shown in Figure 12-7b results from attenuation of the high frequencies. The rounding shown will be either moderate or severe depending on the bandwidth of the amplifier. In other words, by how many harmonics are attenuated by the amplifier frequency response characteristic, and to what degree. This problem becomes especially severe when the fundamental frequency (or pulse repetition rate) is high, the risetime is very fast, and the amplifier bandwidth is low.

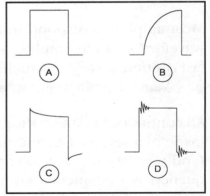

Figure 12-7. Several different ac-response problems.

Frequency compensated operational amplifiers achieve their claimed "unconditional stability" by rolling off the high frequency response drastically above a few kilohertz. A type 741AE is a frequency compensated op-amp with a gain-bandwidth product of 1.25 MHz and an open-loop gain of 250,000. Frequency response at maximum gain is 1.25 MHz/250,000, or 5 kHz. Thus, we can expect good square wave response only at relatively low frequencies. A "rule-of-thumb" for square waves is to make the amplifier bandwidth at least one-hundred times the fundamental frequency. As with all such rules, however, this one should be applied with caution even though it is assimilated into your collection of "standard engineering wisdom."

The other class of problems is shown in Figures 12-7c and 12-7d. In this case we see peaking and ringing of the pulse. Three principal causes of these phenomena are found. First, a skewed bandpass characteristic is where either the low frequencies are attenuated (or amplified less) or the high frequencies are amplified more. Second, there are LC resonances in the circuit that give rise to ringing. Although not generally a problem at the low frequencies produced by most sensors, video operational amplifiers may see this problem. Third, there are both significant harmonics present at frequencies where circuit phase shifts add up to 180 degrees and the loop gain is unity or greater. When combined with the 180 degree phase shift inherent in inverting followers, we have Barkhausen's criteria for oscillation: 360 degree feedback and a loop gain of unity or more. Under some conditions the device will break into sustained oscillation. In other cases, however, oscillation will occur only on fast risetime signal peaks as shown in Figure 12-7d.

## Basic Rules

We must consider several factors when designing inverting follower amplifiers. First, we obviously must consider the voltage gain required by the application. Second, we must consider the input impedance of the circuit. That specification is needed in order to prevent the amplifier input from loading down the driving circuit. In the case of the inverting follower, the input impedance is the value of the input resistor ($R_{in}$), and a simple design rule is in effect:

The input resistor (hence the input impedance) should be equal or greater than ten times the source resistance of the previous circuit.

The implication of this rule is that we must determine the source resistance of the driving circuit, and then make the input impedance (i.e. $R_{in}$ for practical purposes) of the operational amplifier inverting follower at least ten times larger. When the driving source is another operational amplifier we can assume that the source impedance (i.e. the output impedance of the driving op-amp) is 100 ohms or less (it's actually much less). For these cases, make the value of $R_{in}$ at least 1000 ohms (i.e. $10 \times 100$ ohms = 1000 ohms). This value is based on consideration of available driver output load current. In other cases, however, we have a somewhat different problem. Some sensors, for example thermistors for measuring temperature, have a much higher source resistance. One thermistor has an advertised resistance that varies from 10 kohm to 100 kohm over the temperature range of interest, so a minimum input impedance of 1 megohm (i.e. $10 \times 100$ kohm) is required. When the input impedance gets this high, the designer might want to consider the noninverting follower rather than the inverting follower configuration.

In the inverting follower circuit the choice of input impedance drives the design, so is part of the design procedure:

1. Determine the minimum allowable input resistance (i.e. ten or more times the source impedance).

2. If the source resistance is 1000 ohms or less, try 10 kohms as an initial trial input resistance ($R_{in}$).

3. This value might be lowered if the feedback resistor ($R_f$) becomes too high for the required gain. The value of $R_{in}$ is the input resistance, or 10k, whichever is higher.

4. Determine the amount of gain required. In general, the closed-loop gain of a single inverting follower should be less than 500. For gains higher than that figure use a multiple op-amp cascade circuit. Some low-cost op-amps should not be operated at closed-loop gains greater than 200. The reason for this rule is the problems that are found in real (versus ideal) devices. In those cases the distributed gain of a cascade amplifier may prove easier to tame in practical situations.

5. Determine the frequency response (i.e. the frequency at which the gain drops to unity). From steps 3 and 4 we can calculate the minimum gain bandwidth product of the op-amp required.

6. Select the operational amplifier. If the gain is high, e.g. over 100, then you might want to select a BiMOS or BiFET operational amplifier in order to limit the output offset voltage caused by the input bias currents. Select a 741-family device if a) you don't need more than a few kilohertz frequency response, and b) the unconditionally stable characteristics of the 741 is valuable for the application.

7. Also look at the package style. For most applications the 8-pin miniDIP package is probably the easiest to handle. The 8-pin metal can is also useful, and it can be made to fit 8-pin miniDIP positions by correct bending of the leads.

8. Select the value of the feedback resistor:

$$R_f = A_v R_{in}$$

eq. (12-35)

9. If the value of the feedback resistor is too high, i.e. beyond the range of standard values (about 20 megohms or so), or too high for the input bias currents, then try a lower input resistance.

## Altering ac Frequency Response

The natural bandwidth of an amplifier is sometimes too great for certain specific applications. Noise power, for example, is a function of bandwidth as indicated by the expression $P_n = KTBR$. Thus, it is possible that signal-to-noise ratio will suffer in some applications if bandwidth is not limited to that which is actually needed to process the expected waveform. In other cases we find that the rejection of spurious signals suffers if we fail to tailor the bandwidth of an amplifier circuit to that which is required by the bandwidth of the applied input signal. Amplifier stability is improved if the loop gain of the circuit is reduced to less than one at the frequency at which the circuit phase shifts (including internal amplifier phase shift) reaches 180 degrees.

When the distributed phase shift is added to the 180 degrees phase shift seen normally on inverting amplifiers, Barkhausen's criteria for oscillation is satisfied and the amplifier will oscillate. Those criteria are:

1. Total phase shift of 360 degrees at the frequency of oscillation;

2. Output-to-input coupling (may be accidental); and

3. Loop gain of unity or greater.

If these criteria are satisfied at any frequency, then the operational amplifier will oscillate at that frequency. For the present we will discuss just one technique in case you need to know at present.

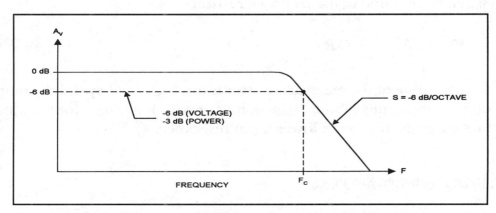

Figure 12-8a. Frequency response with capacitor in feedback loop.

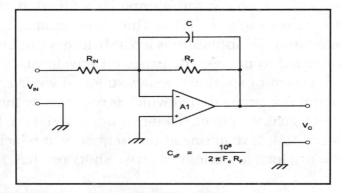

Figure 12-8b. ac-tailored amplifier.

The design goal in tailoring the ac frequency response is to roll off the voltage gain at the frequencies above a certain critical frequency, $F_c$. This frequency is determined by evaluating the application, and is defined as the frequency at which the gain of the circuit drops off -3 dB (-6 dB when voltage is measured) from its in-band voltage gain. The response of the amplifier should look like Figure 12-8a, and is shown here in normalized form in which the maximum in-band gain is taken to be 0 dB. Above the critical frequency the gain drops off -6 dB/octave (an octave is a 2:1 change in frequency) by shunting a capacitor across the feedback resistor, as shown in Figure 12-8b. The reactance of the capacitor is shunted across the resistance of $R_f$, and so reduces the gain. The low-pass filter characteristic is achieved because the capacitive reactance becomes lower as frequency increases. The value of the capacitor is found from:

$$ C = \frac{1}{2\pi R_f F_c} \qquad \text{eq. (12-36)} $$

Where:

C is the capacitance in farads (F).

$R_f$ is the feedback resistance in ohms ($\Omega$).

$F_c$ is the -3 dB frequency in hertz (Hz).

Alternatively, to calculate the capacitance of C in microfarads ($\mu F$) we use Equation 12-37:

$$ C_{\mu F} = \frac{1,000,000}{2\pi R_f F_c} \qquad \text{eq. (12-37)} $$

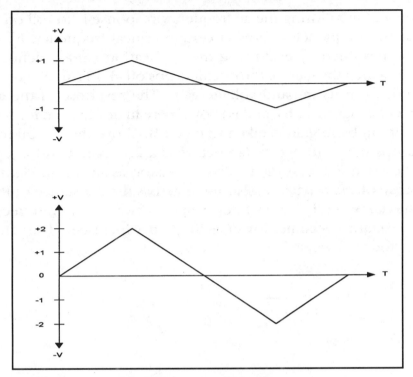

*Figure 12-9. Gain-of-2 noninverting follower.*

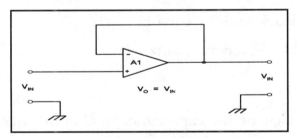

*Figure 12-10. Unity gain noninverting follower.*

## Noninverting Followers

The next standard op-amp circuit configuration is the noninverting follower. This type of amplifier uses the noninverting input of the operational amplifier to apply signal. In this configuration the output signal is in phase with the input signal (Figure 12-9). There are two basic noninverting configurations: unity gain and greater than unity gain.

Figure 12-10 shows the circuit for the unity gain noninverting follower. The output terminal is connected directly to the inverting input, resulting in 100 percent negative feedback. Recall the voltage gain expression for all feedback amplifiers:

$$A_v = \frac{A_{vol} C}{1 + A_{vol} \beta} \qquad \text{eq. (12-38)}$$

Where:

$A_v$ is the closed-loop voltage gain (i.e. gain with feedback).

$A_{vol}$ is the open-loop voltage gain (i.e. gain without feedback).

$\beta$ is the feedback factor.

C is the input attenuation factor.

In this circuit the input signal is applied directly to +IN, so C = 1, and can therefore be ignored. The feedback factor, $\beta$, represents the transfer function of the feedback network. When that network is a resistor voltage divider network, the value of $\beta$ is a decimal fraction that represents the attenuation of the op-amp output voltage before it is applied to the op-amp inverting input. In the unity gain follower circuit the value of $\beta$ is also 1 so it too is ignored. The feedback amplifier equation therefore reduces to:

$$A_v = \frac{A_{vol}}{1 + A_{vol}} \qquad \text{eq. (12-39)}$$

Consider the implications of Equation 12-39 for common operational amplifiers. With a gain of 300,000 (not unusual), Equation 12-39 evaluates to 0.9999967. A gain of 0.9999967 is close enough to 1.0 to justify calling the circuit of Figure 12-10 a "unity" gain follower.

## Applications of Unity Gain Followers

What use is an amplifier that does not amplify? First of all, it is not strictly true that the circuit does not amplify. It has a unity voltage gain, but the power gain is greater than unity. There are three principal uses of the unity gain noninverting follower: buffering, power amplification, and impedance transformation.

A "buffer" amplifier is placed between a circuit and its load in order to improve the isolation between the two. An example is use of a buffer amplifier between an oscillator or waveform generator circuit and its load. The buffer is especially useful where the load exhibits a varying impedance that could result in "pulling" of the oscillator frequency. Such unintentional frequency modulation of the oscillator is very annoying because it makes some oscillator circuits unable to function and causes others to function poorly.

Another common use for buffer amplifiers is isolation of an output connection from the main circuitry of an instrument. An example might be an instrumentation circuit that uses multiple outputs, perhaps one to the A/D converter input of a digital computer, and another to an analog oscilloscope or strip chart recorder. By buffering the analog output to the oscilloscope we prevent short circuits in the display wiring from affecting the signal to the computer, and vice versa.

A special case of buffering is represented by using the unity gain follower as a power driver. A long cable run may attenuate low-power signals. To overcome this problem we sometimes use a low impedance power source to

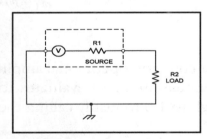

*Figure 12-11a. Equivalent input circuit.*

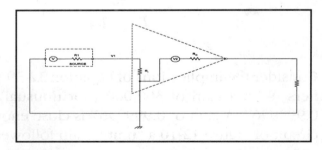

*Figure 12-11b. Equivalent circuit for entire amplifier.*

drive a long cable. This application points out the fact that a "unity" gain follower actually does have power gain (the "unity gain" feature refers only to the voltage gain). If the input impedance is typically much higher than the output impedance, yet $V_o = V_{in}$, then by $V^2/R$ it stands to reason that the delivered power output is much greater than the input power. Thus, the circuit of Figure 12-10 is unity gain for voltage signals and greater than unity gain for power. It is therefore a power amplifier.

The impedance transformation capability is obtained from the fact that an op-amp has a very high input impedance and a very low output impedance. Let's illustrate this application by a practical example. Figure 12-11a is a generic equivalent of a voltage source driving a load (R2). The resistance R1 represents the internal impedance of the signal source impedance. The signal voltage, V, is reduced at the output ($V_o$) by whatever voltage is dropped across source resistance R1. The output voltage is found from:

$$V_o = \frac{V\,R2}{R1\, +\, R2}$$

eq. (12-40)

By way of example: If the ratio of R1/R2 is, say, 10:1, then a 1 Vdc potential is reduced to 0.091 Vdc across R2. Ninety percent of the signal amplitude is lost. With a unity gain noninverting amplifier, as in Figure 12-11b, the situation is entirely changed. If the amplifier input impedance is very much larger

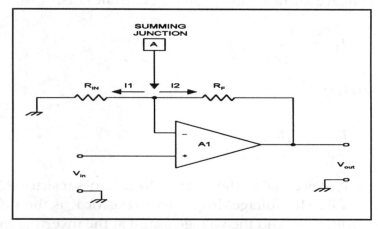

*Figure 12-12a. Noninverting follower with gain.*

than the source resistance, and the amplifier output impedance is very much lower than the load impedance, then there is very little loss and V will closely approximate $V_o$.

## Noninverting Followers with Gain

Figure 12-12a shows the circuit for the noninverting follower with gain. In this circuit, the signal ($V_{in}$) is applied to the noninverting input, while the feedback network ($R_f/R_{in}$) is almost the same as it was in the inverting follower circuit. The difference is that one end of $R_{in}$ is grounded.

We can evaluate this circuit using the same general method as was used in the inverting follower case. We know from Kirchhoff's Current Law, and the fact that the op-amp inputs neither sink nor source current, that I1 and I2 are equal to each other. Thus, the Kirchhoff expression for these currents at the summing junction (point "A") can be written as:

$$I1 = I2 \qquad \text{eq. (12-41)}$$

We know from the properties of the ideal op-amp that any voltage applied to the noninverting input ($V_{in}$) also appears at the inverting input. Therefore:

$$V1 = V_{in} \qquad \text{eq. (12-42)}$$

From Ohm's law we know that the value of current I1 is:

$$I1 = \frac{V1}{R_f} \qquad \text{eq. (12-43)}$$

or, because $V1 = V_{in}$,

$$I1 = \frac{V_{in}}{R_f} \qquad \text{eq. (12-44)}$$

Similarly, current I2 is equal to the voltage drop across resistor $R_f$, divided by the resistance of $R_f$. The voltage drop across resistor $R_f$ is the difference between output voltage $V_o$ and the voltage found at the inverting input, V1. By Ideal Property No. 7, $V1 = V_{in}$. Therefore:

$$I2 = \frac{V_o - V_{in}}{R_f} \qquad \text{eq. (12-45)}$$

We can derived the transfer equation of the noninverting follower by substituting Equations 12-44 and 12-45 into Equation 12-41:

$$\frac{V_{in}}{R_{in}} = \frac{V_o - V_{in}}{R_f} \qquad \text{eq. (12-46)}$$

We must now solve Equation (12-46) for output voltage $V_o$:

$$V_o = V_{in}\left(\frac{R_f}{R_{in}} + 1\right) \qquad \text{eq. (12-47)}$$

Equation 12-47 is the transfer equation for the noninverting follower. The transfer function $V_o/V_{in}$ for a gain-of-2 noninverting amplifier is shown graphically in Figure 12-12b. The solid line assumes no output offset voltage is present (i.e. $V_o = A_v V_{in} + 0$), while the dotted line represents a case where the offset voltage is non-zero.

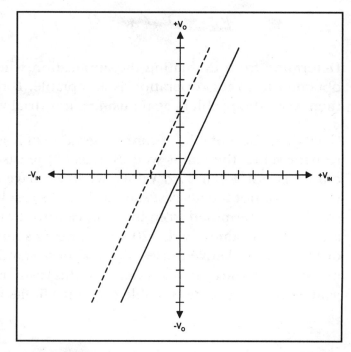

Figure 12-12b.
Effect of offset.

## Advantages of Noninverting Followers

The noninverting follower offers several advantages. In our discussion of the unity gain configuration we mentioned that buffering, power amplification and impedance transformation were advantages. Also, in the gain noninverting amplifier configuration we are able to provide voltage gain with no phase reversal.

The input impedance of the noninverting followers shown thus far is very high, being essentially the input impedance of the op-amp itself. In the ideal device, this impedance is considered infinite, while in practical devices it may range from 500,000 to more than 1012 ohms. Thus, the noninverting follower is useful for amplifying signals from any high impedance source, regardless of whether or not impedance transformation is a circuit requirement.

When the gain required is known (as it usually is in practical situations) we select a trial value for $R_{in}$, and then solve the gain equation to find $R_f$. This new version of the equation is:

$$R_f = R_{in} (A_v - 1) \qquad\qquad eq.\,(12\text{-}48)$$

Determine from evaluating the application whether or not the trial result obtained from this operation is acceptable. If the result is not acceptable, then work the problem again using a new trial value.

What does "acceptable" mean? If the value of $R_f$ is exactly equal to a standard resistor value, then all is well. But, as in the case above, the value (118,800 ohms) is not a standard value. What we have to determine, therefore, is whether or not the nearest standard values result in an acceptable gain error (which is determined from the application). Both 118 kohm and 120 kohm are standard values, with 120 kohms being somewhat easier to obtain from distributor stock inventories. Both of these standard values are less than one percent from the calculated value, so this result is acceptable if a one percent gain error is within reasonable tolerance limits for the application.

## The ac Response of Noninverting Followers

The noninverting amplifier circuits discussed in the preceding sections are dc amplifiers. Nonetheless, as with the inverting amplifiers considered earlier, the noninverting amplifier will also respond to ac signals up to the upper frequency response limit of the circuit.

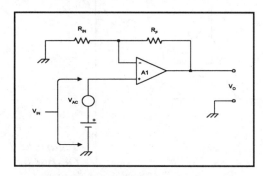

*Figure 12-13.*
*Noninverting follower with*
*DC offset in input signal.*

Figure 12-13 shows the input signal situation for a noninverting follower. In this case there is an ac signal source in series with a dc potential (V1), which are applied to the noninverting input of the operational amplifier. A square wave input signal (Figure 12-14a) is applied to the input, but it is offset by a dc component (Figure 12-14b). If the amplifier has a gain of +2, then the output signal will be as shown in Figure 12-14c. This signal swings from +1 volt to +5 volts. The offset of 1.5 Vdc is amplified by two, and becomes a 3 Vdc offset, with the ac signal swinging about this level.

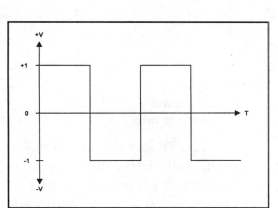

*Figure 12-14a. Symmetrical square signal.*

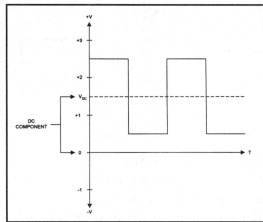

*Figure 12-14b. Square signal with dc component*

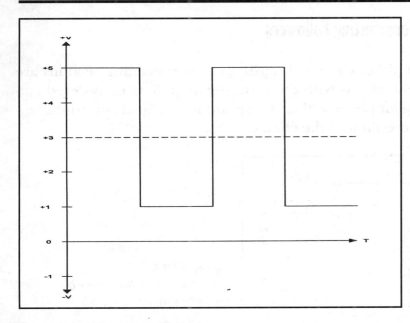

*Figure 12-14c.*
*Output signal.*

## Frequency Response Tailoring

It is possible to custom tailor the upper-end frequency response of the inverting follower operational amplifier with a capacitor shunting the feedback resistor, and the same method also works for the noninverting follower. In this section we will expand the subject and discuss not only tailoring of the upper -3 dB frequency response, but also a lower -3 dB limit as well. The

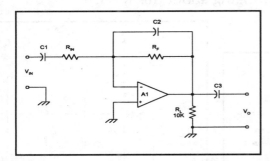

*Figure 12-15.*
*ac-coupled*
*inverting amplifier.*

capacitor across the feedback resistor (Figure 12-15) sets the frequency at which the upper end frequency response falls off -3 dB below the low end in-band gain. The gain at frequencies higher than this -3 dB frequency falls off at a rate of -6 dB/octave (an octave is a 2:1 frequency ratio), or -20 dB/decade (a decade is a 10:1 frequency ratio). The value of capacitor C2 is found by:

$$C2_{\mu F} = \frac{1,000,000}{2\pi R_f F_c} \qquad \text{eq. (12-49)}$$

Where:

$C2_{uF}$ is the capacitance of C2 in microfarads (μF).

$R_f$ is the feedback resistance in ohms (Ω).

$F_c$ is the upper -3 dB frequency in hertz (Hz).

The low frequency response is controlled by placing a capacitor in series with the input resistor, which makes the inverting follower an ac-coupled amplifier. Figure 12-15 is the circuit for an inverting follower that uses ac-coupling at both input and output circuits. Capacitor C2 limits the upper -3 dB frequency response point. Its value is set by the method discussed above. The lower -3 dB point is set by the combination of $R_{in}$ and input capacitor C1. This frequency is set by the equation:

$$C1 = \frac{1,000,000}{2\pi R_{in} F} \qquad \text{eq. (12-50)}$$

Where:

C1 is in microfarads (μF).

$R_{in}$ is in ohms (Ω).

F is the lower -3 dB point in hertz (Hz).

In some cases we will want to ac-couple the output circuit (although it is optional in most cases). Capacitor C3 is used to ac-couple the output, thus preventing any dc component that is present on the op-amp output from affecting the following stages. Resistor $R_L$ is used to keep capacitor C3 from being charged by the offset voltage from op-amp A1. The value of capacitor C3 is set to retain the lower -3 dB point, using the resistance of the stage following as the "R" in the equations above.

Ac-Coupled Noninverting Amplifiers. The noninverting amplifiers discussed thus far have all been dc-coupled. They will respond to signals from either dc or near-dc up to the frequency limit of the amplifier selected. Sometimes, however, we do not want the amplifier to respond to dc or slowly varying near-dc signals. For these applications we select an ac-coupled noninverting follower circuit. In this section we will examine several ac-coupled noninverting amplifiers.

Figure 12-16a shows a capacitor input ac-coupled amplifier circuit. It is essentially the same as the previous circuits, except for the input coupling network. The capacitor in Figure 12-16a serves to block dc and very low frequency ac signals. If the op-amp has zero (or, more realistically, very low) input bias currents, then we can safely delete resistor R3. For all but a few commercially available devices, however, resistor R3 is required if closed-loop gain is high. Input bias currents will charge capacitor C1, creating a voltage offset that is seen by the op-amp as a valid dc signal and amplified to form an output offset voltage. In some devices the output saturates from the C1 charge shortly after turn-on; resistor R3 prevents such latch-ups because it keeps C1 discharged.

Resistor R3 also sets the input impedance of the amplifier. Previous circuits had a very high input impedance because that parameter was determined only by the (extremely high) op-amp input impedance. In Figure 12-16a, however, the input impedance seen by the source is equal to R3.

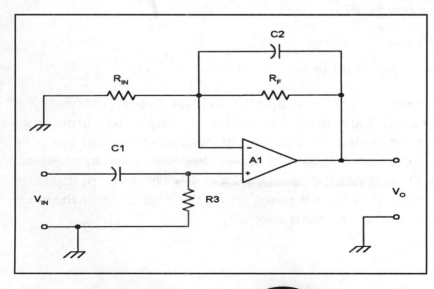

Figure 12-16a.
ac-coupled
noninverting
amplifier.

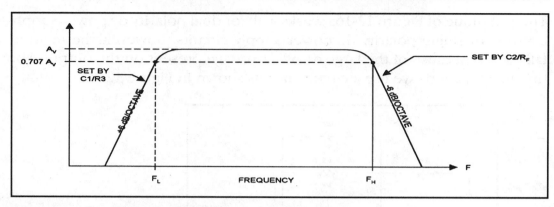

*Figure 12-16b. High-pass filter effect of RC combination.*

Another effect of resistor R3 and capacitor C1 is to limit the low frequency response of the circuit. Filtering occurs because R3C1 forms a high-pass filter (see Figure 12-16b). The -3 dB frequency is found from:

$$F = \frac{1,000,000}{2\pi R1C1}$$  eq. (12-51)

Where:

F is the -3 dB frequency in hertz (Hz).

R1 is the resistance in ohms ($\Omega$).

C1 is the capacitance in microfarads ($\mu$F).

The form of Equation 12-51 is backwards from the point of view of practical circuit design problems. In most cases, we know the required frequency response limit from the application. We also know from the application what the minimum value of R3 should be (derived from source impedance), and often set it as high as possible as a practical matter (e.g. 10 megohms). Thus, we want to solve the equation for C, as shown below:

$$C_{\mu F} = \frac{1,000,000}{2\pi R1F}$$  eq. (12-52)

(all terms as defined above).

The technique of Figure 12-16a works well for dual polarity dc power supply circuits. In single polarity dc power supply circuits, however, the method falls down because of the large dc offset voltage present on the output. For these applications we use a circuit such as shown in Figure 12-17.

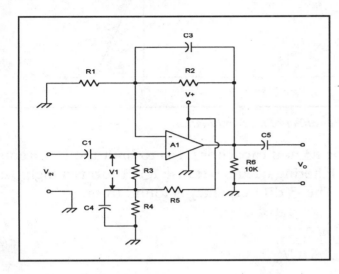

*Figure 12-17. ac amplifier using single dc power supply.*

The circuit in Figure 12-17 is operated from a single V+ dc power supply (the V- terminal of the op-amp is grounded). In order to compensate for the V- supply being grounded, the noninverting input is biased to a potential of:

$$V1 \; = \; (V+)\left(\frac{R4}{R4 \; + \; R5}\right)$$

eq. (12-53)

If R4 = R5, then V1 will be (V+)/2. Because the noninverting input typically sinks very little current, the voltage at both ends of R3 is the same (i.e. V1).

The circuit of Figure 12-17 does not pass dc and some low ac frequencies because of the capacitor coupling. Also, because capacitor C3 shunts feedback resistor R2, there is also a roll-off of the higher frequencies. The high roll-off -3 dB point is found from:

$$F \; = \; \frac{1,000,000}{2\pi R2C3}$$

eq. (12-54)

Where:

F is the -3 dB frequency in hertz (Hz).

R2 is the feedback resistance in ohms ($\Omega$).

C3 is the shunt capacitance in microfarads ($\mu$F).

We can restate Equation 12-54 into a more useful form that takes into account the fact that we usually know the value of R2 (from setting the gain), and the nature of the application sets the minimum value of frequency F. We can rewrite Equation 12-54 in a form that yields the value of C3 from these data:

$$C3 = \frac{1,000,000}{2\pi R3F}$$
eq. (12-55)

The lower -3 dB frequency is set by any or all of several RC combinations within the circuit:

1. R1C2

2. R3C1

3. R3C4

4. $R_L$C5

Resistor R1 is part of the gain-setting feedback network. Capacitor C2 is used to keep the "cold" end of R1 above ground at dc, while keeping it grounded for ac signals.

Resistor R3 is the input resistor and serves the same purpose as the similar resistor in the previous circuit. At midband the input impedance is set by resistor R3, although at the extreme low end of the frequency range the reactance of C4 becomes a significant feature. In general, $X_{C4}$ should be less than or equal to R3/10 at the lowest frequency of operation.

Capacitor C1 is in series with the input signal path and serves to block dc and certain very low frequency ac signals. The value of C1 should be:

$$\bullet \quad C1 = \frac{1,000,000}{2\pi FR3}$$

eq. (12-56)

Where:

C1 is in microfarads (µF).

F is in hertz (Hz).

R3 is in ohms (Ω).

Capacitor C5 is used to keep the dc output offset from affecting the succeeding stage. The 10 kohm output load resistor (R6) keeps C5 from being charged by the dc offset voltage.

# Chapter 13

# Differential and Isolation Amplifiers

The differential amplifier provides two inputs that function together such that the output voltage is proportional to the difference between the signal voltages applied to the two inputs. The reason why operational amplifiers are so often used for differential amplifiers in sensor circuits is that they already have differential inputs, so are naturally adaptable. That is, there are two inputs (-IN and +IN) that each look into the same amount of voltage gain, but are of opposite polarity with respect to each other. The inverting input (-IN) of the operational amplifier provides an output signal that is 180 degrees out of phase with the input signal. In other words, a positive going input signal will produce a negative going output signal, and vice-versa. The noninverting input (+IN) produces an output signal that is in phase with the input signal. For this type of input, a positive going input signal will produce a positive going output signal.

## Input Signal Conventions

Figure 13-1 shows a generic differential amplifier with the standard signals applied. Signals V1 and V2 are single-ended potentials applied to the -IN and +IN inputs, respectively. The differential input signal, $V_d$, is the difference between the two single-ended signals: $V_d = (V2 - V1)$. Signal V3 is a common mode signal, that is, it is applied equally to both -IN and +IN inputs. These signals are described further below.

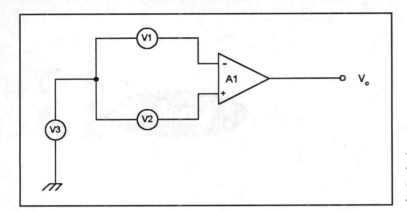

*Figure 13-1.*
*Signals applied*
*to a differential*
*amplifier.*

## Common Mode Signals

First let's consider the common mode signal, V3. A common mode signal is one which is applied to both inputs at the same time. Such a signal might be either a voltage such as V3, or, a case where voltages V1 and V2 are equal to each other and of the same polarity (i.e. V1 = V2). The implication of the common mode signal is that, being applied equally to inverting and noninverting inputs, the output voltage is zero. Because the two inputs have equal but opposite polarity gains for common mode signals, the net output signal in response to a common mode signal is zero.

The operational amplifier with differential inputs cancels common mode signals. An example of the usefulness of this property is in the performance of the differential amplifier with respect to 60 Hz hum pick-up from local ac power lines. Almost all input signal cables for practical amplifiers will pick up 60 Hz radiated energy and convert it to a voltage that is seen by the amplifier as a genuine input signal. In a differential amplifier, however, the 60 Hz field will affect both inverting and noninverting input lines equally, so the 60 Hz artifact signal will disappear in the output.

The practical operational amplifier will not exhibit perfect rejection of common mode signals. A specification called the common mode rejection ratio (CMRR) tells us something of the ability of any given op-amp to reject such signals. The CMRR is usually specified in decibels (dB), and is defined as:

$$CMRR = \frac{A_{vd}}{A_{cm}}$$

*eq. (13-1)*

or, in decibel form:

$$CMRR_{dB} = 20\,LOG\left(\frac{A_{vd}}{A_{cm}}\right)$$

*eq. (13-2)*

Where:

CMRR is the common mode rejection ratio.

$A_{vd}$ is the voltage gain to differential signals.

$A_{cm}$ is the voltage gain to common mode signals.

In general, the higher the CMRR the better the operational amplifier. Typical low-cost devices have CMRR ratings of 60 dB or more, while better devices exhibit CMRR values up to 120 dB.

## Differential Signals

Signals V1 and V2 in Figure 13-1 are single-ended signals. The total differential signal seen by the operational amplifier is the difference between the single-ended signals:

$$V_D = V2 - V1$$

*eq. (13-3)*

The output signal from the differential operational amplifier is the product of the differential voltage gain and the difference between the two input signals (hence the term "differential" amplifier). Thus, the transfer equation for the operational amplifier is:

$$V_o = A_v(V2 - V1)$$

*eq. (13-4)*

That is, the output voltage is the product of the differential gain and the difference between the two input voltages.

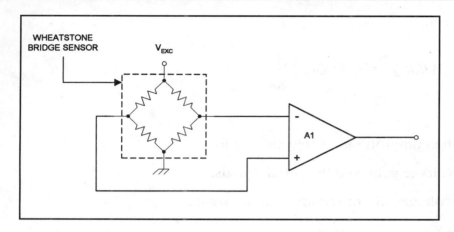

*Figure 13-2. Connection of differential amplifier to bridge sensor.*

## Sensor-Circuit Differential Amplifier Applications

The differential amplifier is used in sensor circuits wherever there are two signals from the same source (e.g. a Wheatstone bridge signal), or a single signal that is balanced with respect to ground (e.g. a magnetometer output signal). Figure 13-2 shows a Wheatstone bridge sensor connected to an excitation source, $V_{EXC}$ and ground, with its two output arms connected to the input of a differential amplifier. The sensor may have all four arms variable, as in a strain gage pressure transducer, or only a single resistor (as in a temperature probe). Note that two of the resistors in the bridge are grounded. It is the signal voltages across these two resistors that form V1 and V2 for the differential amplifier (in terms of Figure 13-1).

## DC Differential Amplifier

The basic circuit for the dc differential amplifier is shown in Figure 13-3. This circuit uses only one operational amplifier, so is the simplest possible configuration. Later you will see additional circuits based on two or three operational amplifier devices. In its most common form the circuit is balanced such that R1 = R2 and R3 = R4. The standard transfer equation for the single op-amp dc differential amplifier is:

$$V_o = V_d \left( \frac{R3}{R1} \right)$$

*eq. (13-5)*

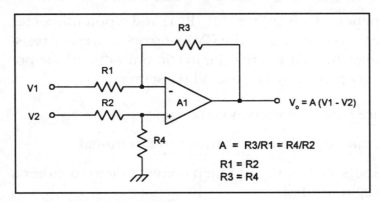

$$A = R3/R1 = R4/R2$$
$$R1 = R2$$
$$R3 = R4$$

*Figure 13-3. Basic dc differential amplifier.*

## Common Mode Rejection

Consider the situation of a differential in the presence of noise signal fields. The noise signal is common mode, so is essentially cancelled by the common mode rejection ratio of the amplifier. Of course, in non-ideal amplifiers the actual situation is that the input signal is subject to the differential gain, while the noise signal is subject to the common mode gain. If an amplifier has a CMRR of 90 dB, for example, the gain seen by the noise signal will be 90 dB down from the differential gain.

The CMRR of the operational amplifier dc differential circuit is dependent principally upon two factors. First, the natural CMRR of the operational amplifier used as the active device. Second, the balance of the resistors, R1 = R2 and R3 = R4. Unfortunately, the balance is typically difficult to obtain with common fixed resistors (not considering precision types). We can use a circuit such as Figure 13-4 to compensate for the mismatch. In this circuit, R1 through R3 are exactly the same as in previous circuits. The fourth resis-

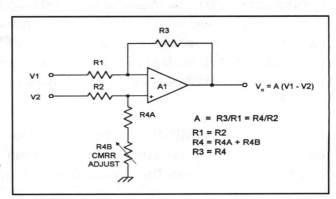

$$A = R3/R1 = R4/R2$$
$$R1 = R2$$
$$R4 = R4A + R4B$$
$$R3 = R4$$

*Figure 13-4. dc differential amplifier with adjustable CMRR.*

tor, however, is a combination of a fixed resistor (R4a) and a potentiometer (R4b). The potentiometer will "adjust out" the CMRR errors caused by resistor and other circuit mismatches. Ordinarily, the maximum value of the potentiometer is ten to twenty percent of the overall resistance.

The adjustment procedure for either version is the same:

1. Connect a zero-center dc voltmeter to the output terminal.

2. Short together inputs V1 and V2, and then connect them to either a signal voltage source, or ground.

3. Adjust potentiometer R4b (CMRR ADJ) for zero volts output.

4. If the output indicator (e.g. a voltmeter or oscilloscope) has several ranges, then switch to a lower range and repeat step 3 above until no further improvement is possible.

Alternatively, connect the output to an audio voltmeter or oscilloscope, and connect the input to a 1 volt to 14 volt peak-to-peak ac signal that is within the frequency range of the particular amplifier. For audio amplifiers, a 400 to 1000 hertz 1 volt signal is typically used.

## Instrumentation Amplifiers

Where a higher gain is required, or the source impedances are higher, then we must resort to a more complex circuit called the op-amp instrumentation amplifier, or "I.A." The simple dc differential amplifier discussed in earlier suffers from several important drawbacks that become important in some applications. First, there is a limit to the input impedance ($Z_{in}$ is approximately equal to the sum of the two input resistors). Second, there is a practical limitation on the gain available from the simple single-device dc differential amplifier. If high gain is attempted, then either the input bias current tends to cause large output offset voltages, or the input impedance becomes too low.

In this section we will demonstrate a solution to these problems in the form of the I.A. All of these amplifiers are differential amplifiers, but offer superior performance over the simple dc differential amplifiers of the last chapter. The instrumentation amplifier can offer higher input impedance, higher gain, and better common mode rejection than the single-device dc differential amplifier.

## Classic I.A. Circuit

The simplest form of I.A. circuit is shown in Figure 13-5. In this circuit the input impedance is improved by connecting inputs of a simple dc differential amplifier (A3) to two input amplifiers (A1 and A2) that are each of the unity gain or higher, noninverting follower configuration (their use here is as buffer amplifiers). The input amplifiers offer an extremely large input impedance (a result of the noninverting configuration) while driving the input resistors of the actual amplifying stage (A3). The overall gain of this circuit is the same as for any simple dc differential amplifier.

It is considered best practice if A1 and A2 are identical operational amplifiers. In fact, it is advisable to use a dual operational amplifier for both A1 and A2 (i.e. two op-amps in a common IC package). The common thermal environment of the dual operational amplifier will reduce thermal drift prob-

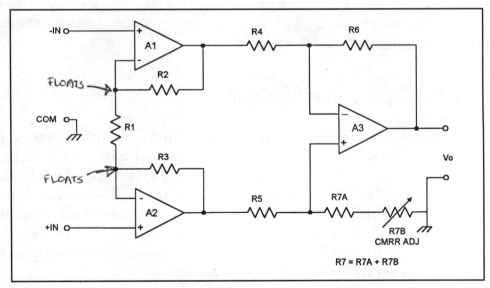

*Figure 13-5. Instrumentation amplifier.*

lems. The very high input impedance of superbeta (Darlington), BiMOS and BiFET operational amplifiers make them ideal for use as the input amplifiers in this type of circuit. The transfer equation for this type of amplifier is:

GAIN OF INPUT STAGE

$$A_v = \left( \frac{2\,R2}{R1} + 1 \right) \left( \frac{R6}{R4} \right) \quad\quad \text{eq. (13-6)}$$

GAIN OF A3 AMP STAGE

Provided: R2 = R3, R4 = R5, and R6 = R7.

## Gain Control for the I.A.

It is difficult to provide a gain control for a simple dc differential amplifier without adding an extra amplifier stage (for example an inverting follower postamplifier with a gain of 0 to -1). For the instrumentation amplifier, however, resistor R1 can be used as a gain control provided that the resistance does not go to a value near zero ohms. This goal can be accomplished with a revised circuit in which resistor R1 is replaced by a series circuit consisting of fixed resistor R1A and potentiometer R1B. This circuit prevents the gain from rising to above the level set by R1A. Don't use a potentiometer alone in this circuit because it can have disastrous effects on the gain. Note in Equation 13-6 that the term R1 appears in the denominator. If the value of R1 gets close to zero, then the gain goes very high (in fact, supposedly to infinity if R1 = 0). The maximum gain of the circuit is controlled by using the fixed resistor in series with the potentiometer.

## Common Mode Rejection Ratio Adjustment

The instrumentation amplifier is no different from any other practical dc differential amplifier in that there will be imperfect balance for common mode signals. The operational amplifiers are not ideally matched, so there will be a gain imbalance. This gain imbalance is further deteriorated by the mismatch of the resistors. The result is that the instrumentation amplifier will respond at least to some extent to common mode signals. As in the simple dc differential amplifier we can provide a common mode rejection ratio adjustment by making resistor R7 variable.

The most popular configuration uses a single potentiometer (R7b) that has a value that is ten to twenty percent larger than the required resistance. For example, if R6 is 100 kohms, then R7 (i.e. R7a and R7b) should be 110 to 120 kohms. Unforunately, these values are somewhat difficult to obtain, so we would pick a standard value for R7b, and then select a value for R7a that is somewhat lower.

## Isolation Amplifiers

An isolation amplifier is one in which there is an extremely high impedance between the ac power mains and the input leads. These amplifiers are typically used in biomedical electronics, where patient safety is an issue, or in situations where high voltage or some other environmental hazard makes it good practice to isolate input from output. Figure 13-6 shows a typical isolation amplifier. It is a differential amplifier (although single-ended isolation amplifiers are available also), with two bipolar dc power supplies. The ±VA supplies are derived from an isolated dc power supply, such as a high frequency carrier-based supply. The ±VB supplies are the normal dc power supplies connected to the ac power mains. Separate grounds are used for both isolated and nonisolated sides.

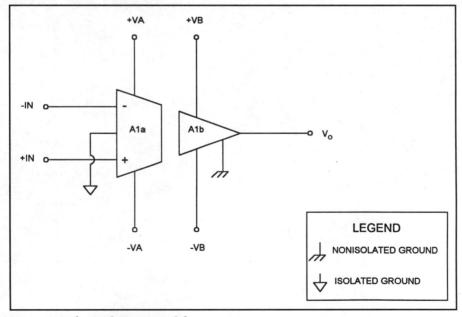

*Figure 13-6. Isolation amplifier.*

# Chapter 14

# Analog Signal Processing Circuits

One of the things that digital computers do quite well is arithmetic. For that reason alone one would assume that analog signal processing circuits, which are essentially analog computers that do mathematical functions, are a thing of the past. Oddly enough, sales of certain special function ICs that do these functions are increasing every year. In this chapter we will look at analog integrators, differentiators, logarithmic and antilog amplifiers and analog multiplier/dividers.

All of the signal processing circuits discussed in this chapter are used in the analog subsystem or "front-end" of computerized sensor-based instruments. Sometimes, when the computer is something like a microcontroller (Chapter 18), the analog implementation is necessary because of the limited repertoire of math functions and memory typically found on those devices. Logarithmic amplifiers are often used to compress signals, or to provide an improvement in dynamic range. Analog multipliers and dividers are often used where sensors need ratiometric treatment.

## Methods for Integration

Integration is the mathematical process of finding the area under a curve. Uses include: 1) acting as a low-pass filter; 2) finding the area under the curve; and 3) finding the time average of a varying waveform (which is re-

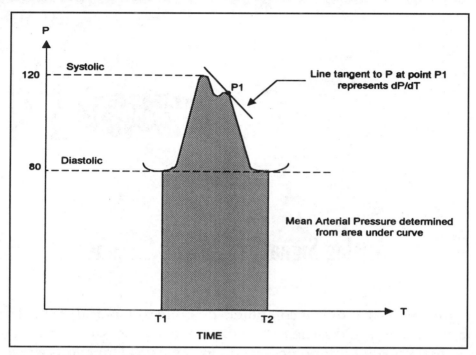

*Figure 14-1. Blood pressure vs. time curve. P is a time-varying voltage from a pressure transducer.*

lated to the second use). If you want to find the area under a time-varying voltage, then you might apply the mathematics of integral calculus to arrive at a number. Alternatively, you might use an analog integrator circuit. The same circuit can also be used to find the time average of a varying voltage.

Time Average. In Figure 14-1 a time-varying voltage signal represents a pressure transducer output. In this particular case, the signal is the output of a human arterial blood pressure transducer used in medical electronics (any other example would also suffice). Notice that the pressure/voltage varies with time from a low non-zero value ($P_D$ or "diastolic") to a high peak value ($P_S$ or "systolic") between times T1 and T2 (which represents one cardiac cycle). If we want to know the mean arterial pressure (MAP), then we would want to find the time average, so we would find the area under the curve and divide by the duration, as shown by:

$$\overline{P} = \frac{1}{t2 - t1} \int_{T1}^{T2} P \, dt + c \qquad \text{eq. (14-1)}$$

Before discussing the circuit, let's first mention the math notation. The $\int$ symbol indicates that the integration process will be applied on the pressure signal, $P$, over the time interval t1 to t2. The $dt$ indicates that the integration takes place with respect to time. The mean arterial pressure over the t2-t1 interval, also denoted by $\overline{P}$, is the integral of the voltage signal representing the pressure. The $c$ is to take care of any offset or constant present. From this illustration we can see that the integrator serves to find the time average "mean" value of an analog voltage waveform - it doesn't have to be a blood pressure signal.

As an aside, I worked for a number of years repairing medical electronic equipment. Many clinical monitors use internal analog or microprocessor-based integrator circuits to find the MAP. Because few nurses know calculus, they are taught a simplification of Equation 14-1 above:

$$\bullet \quad \overline{P} = MAP = P_D + \frac{P_S - P_D}{3} \qquad \text{eq. (14-2)}$$

Equation 14-2 is basically an approximation of Equation 14-1, which is the real equation. The approximation works well, except for many sick people. Unfortunately, many sick people don't exhibit the waveform of Figure 14-1, but rather a damped version in which the notch just prior to t2 is missing, and the pressure simply falls off to the diastolic value. The problem is that this action cuts off about 20 percent of the waveform, which in turn makes the calculated MAP too high. When nurses compare the calculated approximation with the instrumented MAP (which is the actual value!), they almost always trust their training and demand that the instrument be "repaired."

I've told many of them to check the pressure monitoring system with the Calibrate button (a simulated constant pressure), or to disconnect the transducer from the patient and pump a constant pressure (e.g. 100 mmHg) onto the system. If the instrument is working right, a constant pressure for more than about five seconds (the measurement time constant) will yield very nearly identical readings for $P_D$, $P_S$ and MAP. In one case, at 3 AM, an intern had to tell me I was right (and he turned out to have an B.Sc in chemistry, which means lots of calculus).

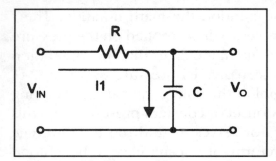

Figure 14-2. Passive resistor-
capacitor (RC) integrator circuit.

## Passive RC Integrator Circuits

Perhaps the simplest form of integrator circuits are made from simple resistor (R) and capacitor (C) elements, as shown in Figure 14-2. You may recognize this circuit as a passive low-pass filter. The RC low-pass filter (integrator) has a -6 dB/octave falling characteristic frequency response.

The operation of the integrator is dependent upon the time constant of the RC network (i.e. R × C). In most cases, we want the integrator time constant to be long (i.e. >10×) compared with the period of the signal being integrated. We can cascade several integrators in order to enhance the effect, and also increase the slope of the frequency response fall off, although only at the expense of severe signal amplitude loss.

## Active Op-Amp Integrator Circuits

The operational amplifier makes it a lot easier to build active integrator circuits. Figure 14-3 shows the standard operational amplifier version of the Miller integrator circuit. An IC operational amplifier is the active element; a resistor (R1) is in series with the inverting input and a capacitor (C1) is in the

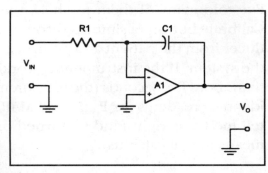

Figure 14-3. Miller
op-amp integrator circuit.

feedback loop. The output voltage of the integrator circuit is dependent upon the input signal amplitude and the RC time constant. The transfer equation for the Miller integrator is:

$$V_O = \frac{-1}{R1C1} \int_0^T V_{in}\, dt + K \qquad\qquad eq.\,(14\text{-}3)$$

Where:

V$_o$ and V$_{in}$ are in the same units (volts, millivolts, etc.).

R1 is in ohms ($\Omega$).

C1 is in farads (F).

t is in seconds (s).

K is a constant in volts (same units as V$_o$ and V$_{in}$).

The expression of Equation 14-3 is a way of saying that the output voltage is equal to the time average of the input signal, plus some constant K which is the voltage that may have been stored in the capacitor from some previous operation (often zero). Alternatively, K may represent an offset error voltage in either the input signal or the operational amplifier itself - and therein is found a problem with textbook integrator circuits.

## Practical Miller Integrator Circuits

The circuit shown in Figure 14-3 is classic, and appears in textbooks (including mine) and magazine articles. Unfortunately, it also doesn't work very well (or at all in some cases) because of the offset voltage problem (the demon K!). These circuits are too simplistic because they depend upon ideal input signals that are symmetrical about zero volts, and the properties of "ideal" operational amplifiers. Unfortunately, the real kind-you-can-go-buy op-amps fall far short of the ideal device that was in the mind of the textbook writer. In real circuits we find that integrators saturate very shortly after turn-on.

The problem with the op-amp integrators was driven home to me when I worked in that hospital lab, and had to build an electronic integrator for one of the customers of our electronics laboratory. When I used a 741 operational amplifier, the output voltage saturated within milliseconds after turn-on. In fact, saturation came so fast that I initially thought the op-amps were bad. The problem was that the input bias currents of the op-amp (which are zero in ideal devices) create a high enough output voltage to fully charge the capacitor in the feedback loop very rapidly.

There is another problem with this kind of circuit, and it magnifies the problem of saturation. This circuit has a very high gain with certain values of R and C. Let's pick an example and see what this gain can mean. The voltage gain $A_v$ of this circuit is given by the term -1/RC, so what is the gain with a 0.01 μF capacitor (certainly not a "large" capacitor in conventional wisdom) and a 10,000 ohm resistor (note: 0.01 μF is $10^{-8}$ farads):

$$A_v = -1/RC$$
$$A_v = -1/(10^4 \text{ ohms})(10^{-8} \text{ farads})$$
$$A_v = -1/10^{-4}$$
$$A_v = -10^4 = -10,000$$

With a gain of -10,000, a +1 volt dc signal applied to the input will try to produce a -10,000 volt output. Unfortunately, the operational amplifier negative output potential is limited to a range of -5 to -20 volts, depending upon the device selected and the applied V- dc power supply voltage. For this case, the operational amplifier will slew to saturation very rapidly! If we want to keep the output voltage from saturating, then we must either keep the RC time constant under control or prevent the input signal from rising too high (not good!). If the maximum output voltage allowable is 10 volts, then the maximum input signal is 10 volts/10,000 or 1 millivolt. Obviously, the best solution is to keep the RC time constant within bounds.

When I built my first analog integrator, and found that 741 devices were not suitable, I turned to the high cost premium grade op-amp devices. At that time, a premium 725 device cost $15, and it suffered the same problems as the 741. The only difference between the $15 premium op-amp and the $0.50 741 device is that on the $15 op-amp the output saturated slowly enough for me to watch it on an oscilloscope or voltmeter - about 4 seconds - instead of nearly instantaneously. Unfortunately, this was still not acceptable. Applying a waveform to the input of even the premium op-amp integrator allowed me to see the output waveform rise up the screen of the oscilloscope and disappear off the top of the screen!

## How to Solve the Problem

Fortunately, there are some practical design tactics that will allow us to keep the integration capability, while getting rid of the problems. A practical integrator is shown in Figure 14-4. The heart of this circuit is a BiMOS operational amplifier, type CA-3140, or its equivalent BiFET type (the CA-3130 or CA-3160 will also work; type CA-3240 is a dual CA-3140 with the same pin-outs as an LM-1458). The reason why this device works so well is that it has a low input bias current (having a MOSFET input circuit with a $1.5 \times 10^{12}$ ohm input impedance). When I tested close to a dozen different op-amps for the circuit, the CA-3140, which cost only about two dollars, out-performed devices costing ten times as much.

Capacitor C1 and resistor R1 in Figure 14-4 form the integration elements, and are used in the transfer equation. Resistor R2 is used both to discharge C1 to prevent dc offsets from either the input signal or the op-amp itself from saturating the device; its value should be 10 to 20 megohms. Resistor

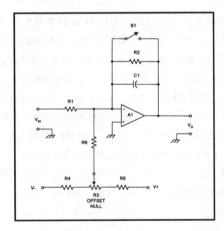

IF R2 IS DELETED, A' 'DC' OR VLF POTENTIAL WILL BUILD ACROSS C1 AND SATURATE $V_o$ TO THE NEGATIVE RAIL.

C2 ⟹ ∞ Ω'S FOR SIGNALS NEAR 'DC'.

*Figure 14-4.*
*Practical Miller*
*op-amp integrator circuit.*

R2 also limits the gain at low frequencies. The RESET switch is used to set the capacitor voltage back to zero (to prevent a "K" factor offset) before the circuit is used. In some measurement applications the circuit initializes by closing S1 (or a relay equivalent) momentarily.

Because of R2 in the circuit we must place a constraint on the transfer equation: the equation is valid only for frequencies greater than or equal to F in Equation 14-4 below:

$$\bullet \quad F_{HZ} \quad = \quad \frac{10^6}{2 \, \pi \, R2 \, C1} \qquad\qquad eq.\ (14\text{-}4)$$

Where:

$F_{Hz}$ is the cut-off frequency in hertz (Hz).

R2 is in ohms ($\Omega$).

C is in microfarads ($\mu$F).

You can also place a compensation resistor (call it $R_c$) between the noninverting input of the operational amplifier and ground. This resistor cancels the effects of input bias current and improves thermal drift performance. It has a value equal to the parallel combination of R1 and R2:

$$\bullet \quad R_c \quad = \quad \frac{R1 \, R2}{R1 \, + \, R2} \qquad\qquad eq.\ (14\text{-}5)$$

There may still be a minor drift problem, so potentiometer R3 is sometimes added to the circuit to cancel it. This component adds a small countercurrent to the inverting input through resistor R6. To adjust this circuit, set R5 initially to mid-range. The potentiometer is adjusted by shorting the $V_{in}$ input to ground (or setting $V_{in} = 0$), and then measuring the output voltage. Press S1 to discharge C1, and note the output voltage should go to zero and stay there. If it does not, then turn R5 in the direction that counters the change of $V_o$ after each time S1 is pressed. Keep pressing S1 and then making small changes in R5 until you find that the output voltage stays very nearly zero, and remains constant after S1 is pressed (there will be some long-term drift normally).

If drift becomes important, and the output voltage range can be limited to less ±5 volts, then it is possible to make the CA-3140 operate in a low-noise mode. Use the 8-pin metal can package type (which are hard to find today), rather than the more common 8-pin miniDIP, and place a expandable heatsink - the kind made for TO-5 metal transistor packages - on it. Limit the dc power supplies to ±5 Vdc.

## Another Method of Integration

Now let's take a look at two methods of integration that are useful under certain other circumstances. One of the methods can be implemented in analog circuitry, but it's a bit difficult. It can also be implemented in micro-computer-based methods of integration. It is called geometric integration. Figure 14-5 shows this special case.

Geometric Integration. This method is usable only in the case where the waveform being integrated is an exponentially decaying voltage. It is used where there is either a high noise level or a significant artifact distorting the waveform somewhere between the 1T and 5T points. The exponential decay waveform is seen a lot in nature, and is also observed when a capacitor discharges through a resistance. By standard convention, the exponential decay is considered completed when five time constants have passed, where one

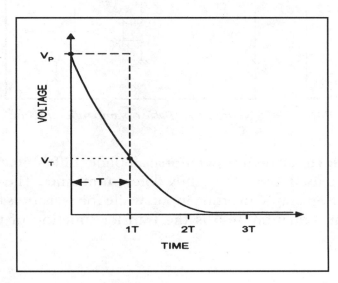

*Figure 14-5.*
*Geometric integration.*

time constant is the time required for the curve to drop to $e^{-1}$ from its peak value (e is the base of the natural logarithms). This curve is difficult to integrate, especially if (as so often happens) there is noise present in the regions later than 1T or 2T.

In geometric integration we create a rectangle with a height equal to the peak voltage of the waveform, and a vertical edge that passes through the 1T point (i.e. point where $V_T = 0.368 \times V_p$). The area of that rectangle is a very good approximation of the area under the exponential decay curve from 0 to 5T. Simply multiply the peak voltage by the time required for 1T to expire, and you will have the area under the curve.

## Differentiation

Figure 14-6 shows the classic RC differentiator circuit. It is similar to the RC integrator, but with the roles of the circuit elements reversed. You will also recognize this circuit as an RC high-pass filter. It produces an output that is proportional to the rate of change of the input signal.

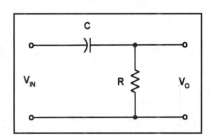

Figure 14-6. RC differentiator.

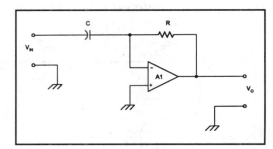

Figure 14-7. Basic op-amp differentiator.

Figure 14-7 shows the basic circuit of the operational amplifier differentiator. Again the RC elements are used, but in a slightly different manner. The capacitor is in series with the op-amp's inverting input, while the resistor is the op-amp feedback resistor. Equation 14-6 is the transfer function of the differentiator.

$$V_o = -RC\frac{dV_{in}}{dt}$$

eq. (14-6)

Where:

V$_o$ and V$_{in}$ are in the same units (volts, millivolts, etc.).

R is in ohms ($\Omega$).

C is in farads (F).

t is in seconds (s).

Equation 14-6 is a mathematical way of saying that output voltage V$_o$ is equal to the product of the RC time constant and the derivative of input voltage V$_{in}$ with respect to time ($\Delta V_{in}/\Delta t$). Since the circuit is essentially a special case of the familiar inverting follower circuit, the output is inverted, hence the negative sign.

Figure 14-8 shows the practical version of the differentiator circuit. The differentiation elements are R1 and C1, and the previous equation for the output voltage is used. Capacitor C2 has a small value (1 pF to 100 pF), and is

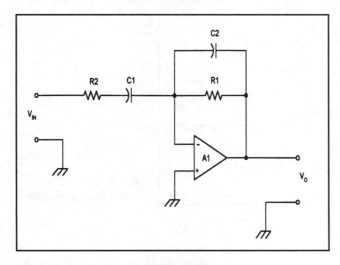

*Figure 14-8.*
*Practical differentiator.*

used to alter the frequency response of the circuit in order to prevent oscillation or ringing on fast risetime input signals. Similarly, a "snubber" resistor (R2) in the input also limits this problem. The operational amplifier can be almost any type with a fast enough slew rate, and the CA-3140 is often recommended. The values of R2 and C2 are often determined by rule-of-thumb; but their justification is taken from the Bode plot of the circuit, and they can be calculated analytically.

## Logarithmic and Antilog Circuits

Logarithmic amplifiers are often used in instrumentation circuits, especially where data compression is required. The overall transfer equation for an operational amplifier circuit is determined by the transfer equation of the feedback network. As might be guessed from this fact, a logarithmic transfer equation can be created in an operational amplifier circuit by placing a non-linear element in the negative feedback loop. An ordinary PN junction transistor meets the requirement. A logarithmic amplifier is one that has a transfer equation of the form either:

$$V_O = K \ Ln \ V_{in} \qquad \qquad eq. (14-7)$$

or,

$$V_o = K \ LOG \ V_{in} \qquad \qquad eq. (14-8)$$

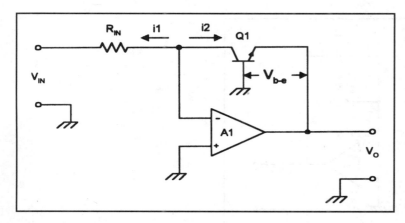

Figure 14-9. Basic logarithmic amplifier.

Figure 14-9 shows the basic circuit for an inverting logarithmic amplifier. As with any inverting amplifier, we can assume that the summing junction potential is zero because the noninverting input is grounded. In basic transistor theory the base-emitter voltage of the transistor is given by:

$$V_{b-e} = \frac{KT}{q} Ln\left(\frac{I_c}{Is}\right) \qquad \qquad eq. (14-9)$$

Where:

$V_{b-e}$ is the base-emitter potential is volts (V).

K is Boltzmann's constant ($1.38 \times 10\text{-}23$ J/K).

T is the temperature in degrees Kelvin (K).

q is the electronic charge ($1.6 \times 10\text{-}19$ coulombs).

Ln indicates the natural (or "base-e") logarithms.

$I_c$ is the collector current of the transistor in amperes (A).

$I_s$ is the reverse saturation current of the transistor (approximately $10^{-13}$ amperes at 300 K).

Because the configuration of Figure 14-9 makes $V_{b-e} = V_o$:

$$V_o = \frac{KT}{q} \operatorname{Ln}\left(\frac{I_c}{I_s}\right)$$
eq. (14-10)

or, for those who prefer base-10 logarithms:

$$\frac{\text{LOG } X}{\text{Ln } X} = 0.4343$$
eq. (14-11)

and,

$$\frac{\text{Ln } X}{\text{LOG } X} = 2.3$$
eq. (14-12)

so,

$$V_o = \frac{2.3KT}{q} \operatorname{LOG}\left(\frac{I_c}{I_s}\right)$$
eq. (14-13)

$$V_o \approx 60\, mV \operatorname{LOG}\left(\frac{I_c}{I_s}\right)$$
eq. (14-14)

The equation above demonstrates that the output voltage $V_o$ is a logarithmic function. From KCL it is known that:

- $$I1 = -I_c$$ 

eq. (14-15)

And from Ohm's law:

- $$I1 = \frac{V_{in}}{R_{in}}$$

eq. (14-16)

Substituting Equations 14-15 and 14-16 into Equation 14-14 yields:

- $$V_o = 60\,mV\,LOG\left(\frac{I1}{I_s}\right)$$

eq. (14-17)

or,

- $$V_o = 60\,mV\,LOG\left(\frac{V_{in}/R_{in}}{I_s}\right)$$

eq. (14-18)

Thus, the output voltage $V_o$ is proportional to the logarithm of the input voltage $V_{in}$. The simple circuit of Figure 14-9 is the one usually published in textbooks, but in a practical sense it only works some of the time due to the realities of non-ideal operational amplifiers. But there is one common problem: spurious oscillation. Unfortunately, the amplifier often oscillates in the basic configuration.

A modified version of the logarithmic amplifier circuit is shown in Figure 14-10. In this circuit a compensation network (R2/R4/C1) is added to prevent the oscillation. The values of the network components (except R4) is found from empirical data based on the following approximations:

- $$R2 = \frac{v_{o(max)} - 0.7}{(V_{in(max)}/R_1) + (V_{o(max)}/R4)}$$

eq. (14-19)

and,

- $$C1 = \frac{1}{2\,\pi\,F\,R_3}$$

eq. (14-20)

Another problem of the logarithmic amplifier is temperature sensitivity. Recall that the operant equation for the logarithmic amplifier is:

- $$V_o = \frac{KT}{q}\,Ln\left(\frac{V_{in}/R_{in}}{I_s}\right)$$

eq. (14-21)

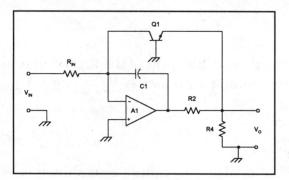

*Figure 14-10. Frequency compensated log amp.*

The "T" term in the equation is temperature in degrees Kelvin. Temperature is a variable rather than a constant, so we can expect the output voltage to be a function of both the applied input signal voltage and the temperature of the b-e junction in the transistor. This temperature is, in turn, a function of ambient temperature. In order to prevent pollution of the output signal data it is necessary to temperature compensate the logarithmic amplifier circuit. Figure 14-11 shows an approach to the temperature compensation job. The value of R2 in Figure 14-11 is approximately 15.7 times the temperature of the thermistor ($R_t$) at room temperature.

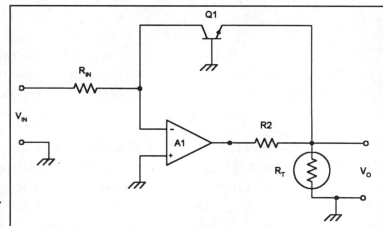

*Figure 14-11. Temperature compensated log amp.*

## Antilog Amplifiers

The antilog amplifier performs the inverse function of the logarithmic amplifier. The output voltage from the antilog amplifier is $V_o = $ Log-1 $V_{in}$, or in terms of Figure 14-12:

$$V_o = \frac{I_s\, e^{qV_{in}/KT}}{R1}$$

eq. (14-22)

The simplified circuit for a conventional antilog amplifier is shown in Figure 14-12. Note that, again, a PN junction from a bipolar transistor is used because of its logarithmic transfer function. But note here that the respective positions of the transistor and resistor are reversed from the logarithmic amplifier.

## Analog Multiplier/Divider Circuits

An analog multiplier and divider (Figure 14-13) is a special function circuit that produces an output voltage that is proportional to either the product or the quotient of two input voltages ($V_X$ and $V_Y$):

$$V_o = \frac{V_X}{KV_Y}$$

eq. (14-23)

Or,

$$V_o = KV_X V_Y$$

eq. (14-24)

These circuits come in two principle varieties: two-quadrant and four-quadrant. These designations refer to whether or not the output is positive only (two-quadrant) or can be positive or negative (four-quadrant) depending on the state of the input voltages.

Figure 14-14 shows the use of a divider circuit as a ratiometric sensor interface. A ratiometric device is one where the output is proportional to the quotient of the two input voltages. In this case, drift in the excitation potential would normally produce artifacts in the output signal. But when the excitation potential and the sensor output signal are applied to the divider, the output will be a signal voltage that is independent of the excitation potential variation.

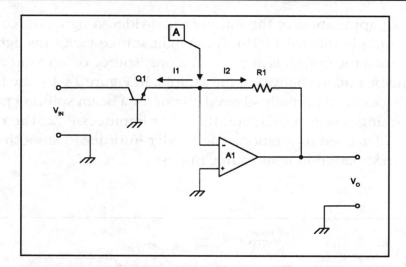

*Figure 14-12 . Antilog amplifier.*

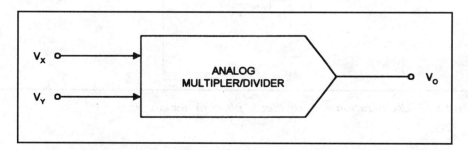

*Figure 14-13.  Analog multiplier/divider.*

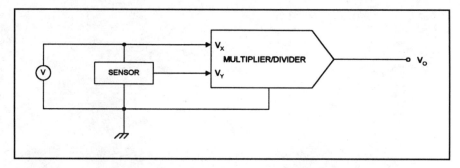

*Figure 14-14. Analog divider used for ratiometric sensors.*

Consider an application of the ratiometric divider. A light source is passed through a sample (Figure 14-15). If the light source varies (as light sources often do), then the output is in error. The only source of light intensity variation should be due to changes in the sample. In Figure 14-15, the light beam is split using either a partially silvered mirror, or a beam splitting prism. One beam is impinges sensor PC1, and the other impinges PC2. The outputs of PC1 and PC2 are fed to a ratiometric divider in order to smooth out variations due to variation in light source intensity.

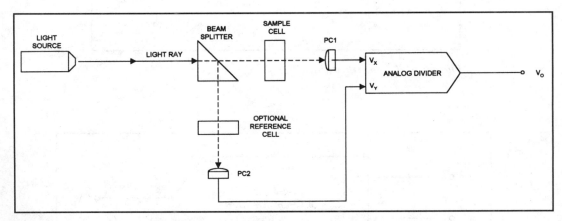

*Figure 14-15. Use of ratiometric divider in photometer circuit.*

# Chapter 15

# Removing 60 Hz Hum from Sensor Circuits

The AC power lines that bring us the benefits of electrical power also cause a tremendous amount of interference in sensor based instruments. Those 60 Hz artifacts can completely obliterate the actual data being sought, especially when the signal is weak. For example, when measuring the human electro-cardiogram, the desired signal has a 1 $\mu$V peak, while the 60 Hz interference could have several millivolts peak amplitude. Similarly, 60 Hz affects opto-electronic sensors, some magnetic sensors and most piezoresistive strain gages.

The best way to deal with 60 Hz interference is to prevent it from happening in the first place. This chapter deals with some methods for making that happen. But what happens when 60 Hz already exists, or is unavoidable? Also, what happens when some other frequency (e.g. the 400 Hz carrier in a pressure amplifier) interferes with the data? The answers are discussed in this chapter.

Frequency selective filters are used to remove unwanted frequencies from electronic circuits. Most readers are familiar with such filters as the low-pass filter, the high-pass filter and the bandpass filter. These filters will remove whole bands of frequencies; for example, the low-pass filter attenuates all frequencies above the cut-off frequency. But what do you do for signals that are on a frequency that is within the band of interest?

A good example of an unwanted in-band signal is the 60 Hz hum pick-up from the power lines in your home and neighborhood. Industrial, scientific and medical instrumentation that uses a frequency band under 500 Hz or 1000 Hz, which encompasses 60 Hz, would be quite susceptible to interference. Figure 15-1 shows an example of 60 Hz artifact linearly mixed with a low frequency signal, and recorded on an oscilloscope. Other forms of interference show a band of hum riding on the signal.

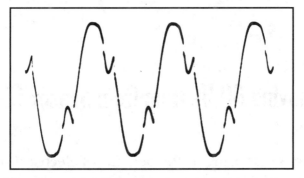

*Figure 15-1. 60 Hz interference to an electronic signal.*

Other examples of interfering in-band signals might come when oscillators, motors, or other signal generating signals are used in the circuit (or nearby circuits). These signal sources are often difficult to suppress, especially if they are of high power or errors were made in laying out the circuit. Frequencies from 100 Hz to 25 kHz are commonly found, although others are also possible.

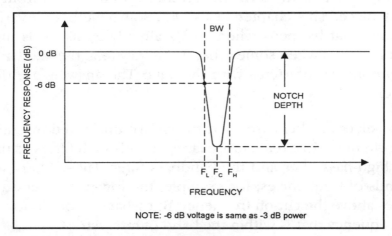

*Figure 15-2. Notch filter response curve.*

## Notch Filters

The solution to unwanted in-band frequencies is the notch filter. The frequency response of a notch filter is shown in Figure 15-2. These filters are similar to another class, i.e. bandstop filters, but the band of rejection is very narrow around the center frequency ($F_c$). The bandwidth (BW) of these filters is the difference between the frequencies at the two -6 dB[3] points, when the out-of-notch response is the reference 0 dB point. These frequencies are $F_L$ and $F_H$, so the bandwidth is $F_H - F_L$.

The "sharpness" of the notch filter is a measure of the narrowness of the bandwidth, and is specified by the "Q" of the filter. The Q is defined as the ratio of the center frequency Fc to bandwidth:

$$Q = \frac{F_c}{BW} \qquad\qquad eq.\ (15\text{-}1)$$

For example, a notch filter that is centered on 60 Hz, and has -6 dB points of 58 and 62 Hz (4 Hz bandwidth) has a Q of 60/4 or 15.

The notch filter does not remove the entire offending signal, but rather suppresses it by a large amount. The notch depth (see Figure 15-2) defines the degree of suppression, and is defined by the ratio of the gain of the circuit at an out-of-notch frequency (e.g. $F_{ob}$) to the gain at the notch frequency. Assuming equal input signal levels at both frequencies (which has to be checked, most signal generators have variable output levels with changes of frequency!), the notch depth can be calculated from the output voltages of the filter at the two different frequencies:

$$Notch\ Depth = 20\ \text{LOG}\left(\frac{V_{fc}}{V_{ob}}\right) \qquad\qquad eq.\ (15\text{-}2)$$

Notch depths of -40 to -60 dB are relatively easy to obtain with proper circuit design and component selection.

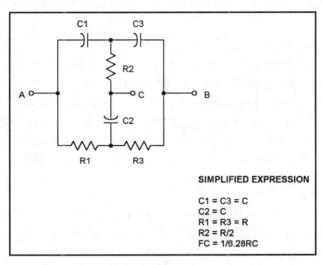

**SIMPLIFIED EXPRESSION**

C1 = C3 = C
C2 = C
R1 = R3 = R
R2 = R/2
FC = 1/6.28RC

*Figure 15-3. Twin-tee passive RC notch filter.*

## Twin-Tee Filter Networks

One of the most popular forms of notch filter is the twin-tee filter network, shown in Figure 15-3. It consists of two T-networks, consisting of C1/C3/R2 and R1/R3/C2. The center notch frequency of the network in the generic case is given by:

$$F_c \;=\; \frac{1}{2\pi} \sqrt{\frac{C1 + C3}{C1\,C2\,C3\,R1\,R3}} \qquad\qquad eq.\,(15\text{-}3)$$

We can simplify the expression above by adopting a convention that calls for the following relationships:

C1 = C3 = C

R1 = R3 = R

C2 = C

R2 = R/2

If this convention is adopted, then we can reduce the frequency equation to:

$$F_c \;=\; \frac{1}{2\pi R C} \qquad\qquad eq.\,(15\text{-}4)$$

In these expressions, F is in hertz (Hz), R is in ohms and C is in farads. Be sure to use the right units when working these problems: "10 kohms" is 10,000 ohms, and "0.001 µF" is $1 \times 10^{-9}$ farads. In calculating values, it is usually prudent to select a capacitor value, and then calculate the resistance needed. This is done for two reasons: 1) there are many more standard resistance values, and; 2) potentiometers can be easily used to trim the values of resistances, but it is more difficult to use trimmer capacitors. For 60 Hz filters, some common values for R and C are:

| C | R |
|---|---|
| 0.001 µF | 2,652,582 Ω |
| 0.01 µF | 5,258 Ω |
| 0.15 µF | 17,684 Ω |

One of the problems of these filters is that the depth of the notch is a function of two factors involving these components. First, that they are very close to the calculated values, and second, that they be matched closely together. For example, a 60 Hz notch filter was built using the 0.15 µF and 17,684 Ω values from the table above. The 0.15 µF capacitors were selected at random from a group of a dozen or so "mine run" capacitors of good quality, while the resistors were 18 kohm, 5% metal film resistors. The notch depth at 60 Hz was only 10 dB, but at 58 Hz it was 48 dB. The mismatch caused a significant shift of notch frequency.

A second filter was built using the same values. In this case, the 0.15 µF capacitors were selected from about 20 on hand (precision components are difficult to obtain). In order to match them as close as possible, the capacitance of each was measured using the capacitance tester function on a low-cost (< $100) digital multimeter. The order of priority of selection was to find those that closely matched each other, and only incidentally how close they come to the calculated value. Errors in the mean capacitance of the selected group can be trimmed out using a potentiometer in the resistor elements of the twin-tee network.

Figure 15-4 shows a 60 Hz twin-tee notch filter that uses potentiometers to tune the filter's central frequency. When adjusting this circuit, be sure to perform each potentiometer adjustment several times, for the adjustments are very interactive. Connect a signal source at the input of the circuit, and monitor the output on an oscilloscope or ac voltmeter. Adjust R1B, R2B and R3B several times each to null the output signal. Stop when there is no further suppression of the output signal.

When selecting a frequency source, either select a well-calibrated source, or use a frequency counter to measure the frequency. Keep in mind the situation described above where only a 2 Hz shift produced a 38 dB difference in notch depth! Alternatively, use a 6.3 volt or 12.6 volt ac filament transformer secondary as the signal source (WARNING: The primary circuits of these transformers are at a potential of 115 volt ac, and can thus be lethal if mishandled!)

Another form of adjustable circuit is shown in Figure 15-5. It is similar in philosophy to Figure 15-4, but uses variable trimmer capacitors rather than potentiometers as the adjustable element. In this circuit, C = 265 pF and R = 10 meg$\Omega$. The selected values of capacitance combine a fixed capacitor and a variable capacitor in order to be able to adjust above and below the design value by a margin sufficient to cover any tolerance problems. The 10 meg$\Omega$ resistors should be matched using an ohmmeter. Again, the goal is to match the resistances close to each other, and only incidentally close to the

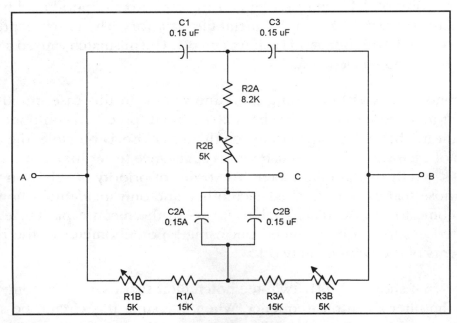

*Figure 15-4. Variable notch filter using potentiometers.*

design value. Once again, it is desirable to have the resistance values "right on," but failing that, they should be grouped as close together as possible, with the difference between their mean value and the design value being trimmed out by the variable capacitances.

## Active Twin-Tee Notch Filters

Active frequency selective filters use an active device such as an operational amplifier to implement the filter. In the active filter circuits to follow, the "twin-tee" networks are shown as block diagrams for sake of simplicity, and are identical to those circuits shown earlier; the ports "A," "B," and "C" in the following circuits are the same as in the previous networks.

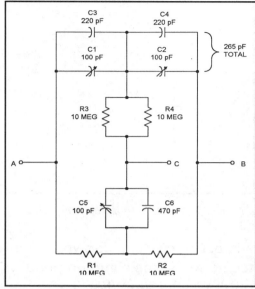

*Figure 15-5. Variable notch filter using variable capacitors.*

The simplest case of a twin-tee filter is to simply use it "as is," i.e. use the filter circuits shown above. But the better solution is to include the twin-tee filter in conjunction with one or more operational amplifiers. Figure 15-6 shows the basic application in which the twin-tee network is cascaded with an input buffer amplifier (optional) and an output buffer amplifier (required). The purpose of buffer amplifiers is to isolate the network from the outside world. In both cases A1 and A2, the buffer amplifiers are operational amplifiers connected in a unity gain, noninverting follower circuit. For low frequency applications, the op-amps can be 741, 1458 and other similar devices. For higher frequency applications, i.e. those with an upper cut-off frequency above 3 kHz, use a non-frequency compensated device such as the CA-3130 or CA-3140 devices.

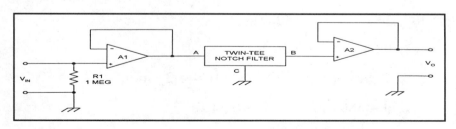

*Figure 15-6. Buffered notch filter.*

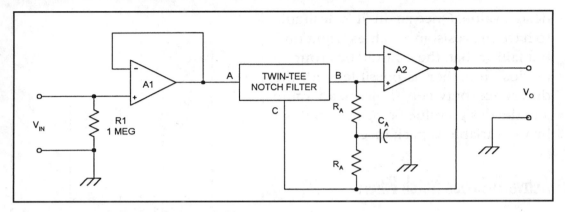

*Figure 15-7. Active twin-tee notch filter.*

A superior circuit is shown in Figure 15-7. In this circuit, port C of the twin-tee network (the common point) is connected to the output terminal of the output buffer amplifier. There is also a feedback network consisting of two resistors ($R_a$) and a capacitor ($C_a$). The values of R and C in the twin-tee network are found from Equation 15-4 above, while the values of $R_a$ and $C_a$ are found from:

$$R_a = 2RQ \qquad\qquad eq. (15\text{-}5)$$

and,

$$C_a = \frac{C}{Q} \qquad\qquad eq. (15\text{-}6)$$

Example

Design a 60 Hz notch filter with a Q of 8.

1. Select a trial value for C: 0.01 µF.

2. Calculate the value of R from Equation 15-4: 265,392 Ω.

3. Calculate R/2: 265,392/2 = 132,696 Ω.

4. C2 = C = (2)(0.01 µF) = 0.02 µF.

5. Select $R_a$: $R_a$ = 2QR = (2)(8)(265,392 Ω) = 4.24 megΩ.

6. Select $C_a$ = C/Q = 0.01 µF /8 = 0.0013 µF.

When Figure 15-7 was built using the twin-tee network of Figure 15-4, with the potentiometers for adjustment, the null was close to -48 dB deep. Figure 15-8 shows the input (upper) and output (lower) traces for a signal that was on a frequency of precisely 60 Hz.

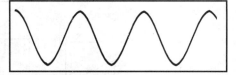

*15-8 Top trace is 60 Hz sine wave signal at input of filter, bottom trace (to same scale) is filter output.*

Figures 15-9a and 15-9b show two variations on the circuit that may be preferred by some people. The circuit in Figure 15-9a shows the addition of a notch depth control (R2), consisting of a 5 kohm potentiometer in a feedback loop between the output of amplifier A2 and the input port of the twin-tee network. A variable Q control is shown in Figure 15-9b. In this circuit, a

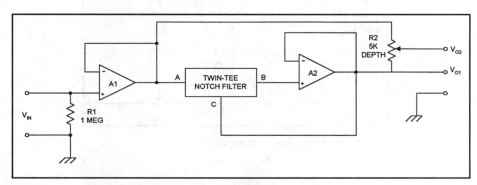

*Figure 15-9a. Active notch filter with tunable depth.*

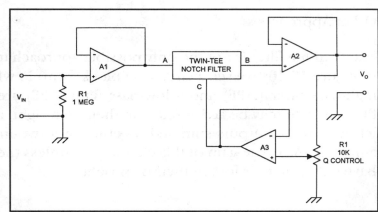

*Figure 15-9b. Better version.*

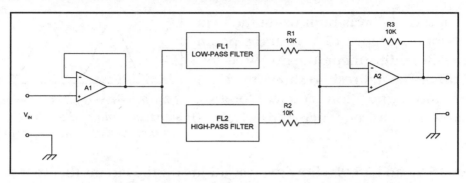

*Figure 15-10a. Using matched LPF and HPF filters to obtain a notch.*

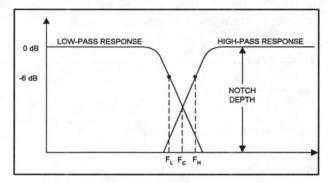

*Figure 15-10b. Response curves.*

noninverting follower (A3) is connected in the feedback loop in place of $R_a$ and $C_a$. The Q of the notch is set by the position of the 10 kohm potentiometer (R1). Values of Q from 1 to 50 are available from this circuit.

Other Approaches

The twin-tee filter is not the only possible approach to making a notch filter. In Figure 15-10a we see a circuit (in block form) in which the responses of a high-pass filter (HPF) and a low-pass filter (LPF) are overlapped. The two filters, which may be active filters in their own right, are connected in parallel, with their outputs summed together in a two-input inverting follower amplifier (A2). The gain of this circuit is -1, unless the HPF and LPF sections have either gain or loss in their own right.

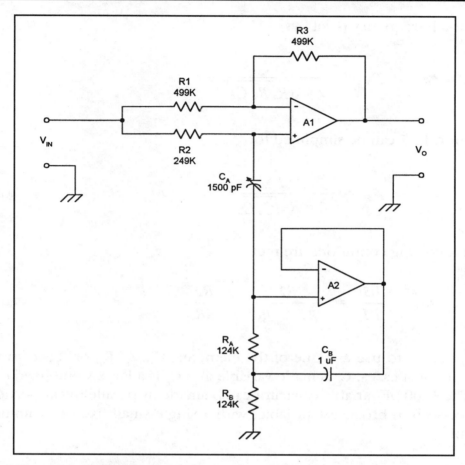

*Figure 15-11. Gyrator notch filter.*

The overlapped frequency responses are shown in Figure 15-10b. The lower notch frequency ($F_L$) is the -6 dB frequency of the LPF, while the upper notch frequency ($F_H$) is the -6 dB frequency of the HPF. The depth of the notch is set by the point where the responses of the LPF and HPF touch.

The notch filter of Figure 15-10a is often implemented using a single state variable filter circuit because those types of active filters have both low-pass and high-pass output ports.

Another approach to notch filter circuits is shown in Figure 15-11. This circuit is sometimes called the gyrator or active inductor notch filter (it's also sometimes called the virtual inductor notch filter).

The notch frequency is set by:

$$\bullet \qquad F_c \;=\; \frac{1}{2\pi\sqrt{R_a\,R_b\,C_a\,C_b}} \qquad\qquad eq.\ (15\text{-}7)$$

Equation 15-7 can be simplified to:

$$\bullet \qquad F_c \;=\; \frac{1}{2\pi R\sqrt{C_a\,C_b}} \qquad\qquad eq.\ (15\text{-}8)$$

if the following conditions are met:

$$\bullet \qquad \frac{R3}{R1} \;=\; \frac{R2}{R_a + R_b} \;=\; \frac{R2}{2R} \qquad\qquad eq.\ (15\text{-}9)$$

It is possible to use any one of the elements, $C_a$, $C_b$, $R_a$, or $R_b$, to tune the filter. In most cases, $C_a$ is made variable and $C_b$ is a large value fixed capacitor. The 1500 pF variable capacitor can be made by paralleling all sections of a three-section broadcast variable, with a single small fixed or trimmer capacitor.

## Conclusion

Unwanted signals, such as 60 Hz interference, are a pain in the transistor, but they are also quite easy to get rid of using the ordinary notch filter circuits shown above.

# Chapter 16

# Instrumentation Circuit Shielding, Grounding Input Control

It is common for books on instrumentation to cover amplifier and sensor circuits, but that information is insufficient to provide profound knowledge of the subject. Some of the often overlooked matters can make a tremendous difference in the way an instrumentation system works. In this chapter, we will take a look at amplifier input control circuits, input shielding, circuit shielding and filtering for electromagnetic interference control, and grounding techniques. This seems to be a duchesses mixture of topics, but, as you will come to understand, they are intertwined to a great extent.

## Input Control

The basic amplifier input configurations are shown in Figure 16-1. Although the noninverting input is shown here, similar circuits are used for the inverting circuit. Figure 16-1a shows the basic direct coupled input connection. It passes signals from dc to the maximum bandwidth of the amplifier. The input connector center conductor is connected directly to the amplifier input. In many cases, an input resistor is connected across the input circuit (typically it has a value of 1 to 10 megΩ).

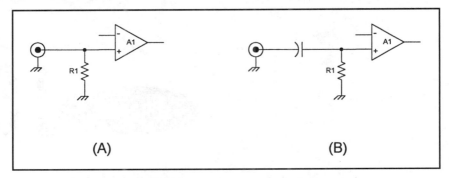

(A)                                      (B)

*Figure 16-1 a) dc-coupled input; b) ac-coupled input.*

A typical ac coupled input is shown in Figure 16-1b. This circuit is similar to the direct couple circuit, except that a capacitor is added in series with the input line. The resistor (R1) is not optional except when the amplifier has an extremely high input impedance (small input bias currents tend to charge capacitors, and the resistor provides a discharge path).

The capacitor blocks any dc component in the input signal, while the resistor and capacitor together determines the lower -3 dB point in the frequency response characteristic. This frequency is:

$$F_{-3dB} = \frac{10^6}{2\pi R1 C}$$

eq. (16-1)

Where:

$F_{-3dB}$ is the lower -3 dB point in hertz (Hz).

R1 is in ohms ($\Omega$).

C is in microfarads ($\mu$F).

The amplifier circuits presented above have a relatively limited use because of the lack of input circuitry control. In this section we will take a look at a couple of ways that the input circuitry can be controlled to provide both ac and dc coupling in the same amplifier, as well as selectable input voltage range.

Figure 16-2 shows two methods for obtaining the ac-GND-dc coupling system often seen in oscilloscopes and other instruments. In both cases, assume that the basic input amplifiers are dc-coupled. Also, the differential amplifier case is shown here, but the same ideas are also used on single-ended amplifiers. Figure 16-2a shows the use of a rotary switch input control circuit, while Figure 16-2b shows a slide switch version (the latter is used more often).

Resistors R1 and R2 in Figure 16-2a are 10 megΩ each, with 1% or better tolerance. In some cases, where extremely high input impedance amplifiers are used, these resistors are not needed. If in doubt, use them, although keep in mind that any mismatch in their values will deteriorate the CMRR of the overall circuit.

Switch S1A/B is a two pole, three throw (2P3T) rotary switch mounted on the front panel. In position "A" the inputs of the amplifiers (A1 and A2) are connected directly to the input jacks (J1 and J2). In this position the amplifier is dc coupled. In position "B", however, the amplifier inputs are disconnected from the input jacks and then are grounded. This "GND" position can be used to check or set the zero baseline control, or to find zero on the external oscilloscope or A/D converter input. In position "C" the signals from the input jacks (J1 and J2) are coupled through capacitors C1 and C2, respectively.

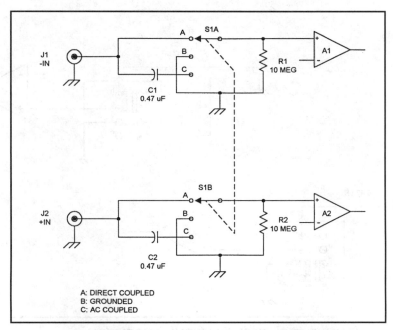

*Figure 16-2 a. Input switching to provide ac-GND-dc function.*

The input control circuit of Figure 16-2b uses a two pole, three position (2P3P) slide switch. In position 1, terminals a-b and g-h are shorted by the internal sliders; in position 2 terminals b-c and f-g are shorted; and in position 3 terminals c-d and e-f are shorted. Position no. 1 is ac-coupled. Signal applied to the input terminal passes through capacitor C1 and terminals g-h of S1 directly to the noninverting input of A1. In position no. 2, the amplifier input is grounded through terminals f-g. The signal input jack is disconnected from the amplifier at this point. This position is often used to establish the baseline or zero line. In position no. 3, signal applied to the input flows through terminals c-d of S1 through a wire to terminal g and then to the noninverting input of A1.

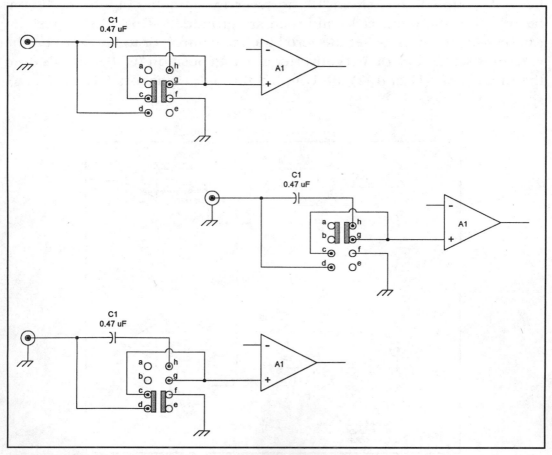

*Figure 16-2b.  Slide switch implementation for all three modes.*

Figure 16-3 shows a frequency compensated voltage divider input circuit that permits switch selectable input ranges to be accommodated. Also provided are signal overload clipping diodes (D1 and D2) to protect the input stages of the amplifier devices. Resistors R1, R2, R3 and the parallel combination of R4/R5 form a resistive voltage divider with each position of selector switch S1 being 10:1 relative to the next lower order position. Thus, if position "A" is 100 µV full-scale, then "B" is 1.00 volt full-scale, "C" is 10.00 volts full-scale and "D" is 100 volts full-scale.

The frequency compensation is accomplished by capacitors shunting the resistors in the voltage divider network. These are adjusted using a square wave (1000 Hz, unless this frequency is outside the amplifier passband) input signal. While monitoring the signal on an oscilloscope, adjust each trimmer in turn for the best "squareness" for that particular setting.

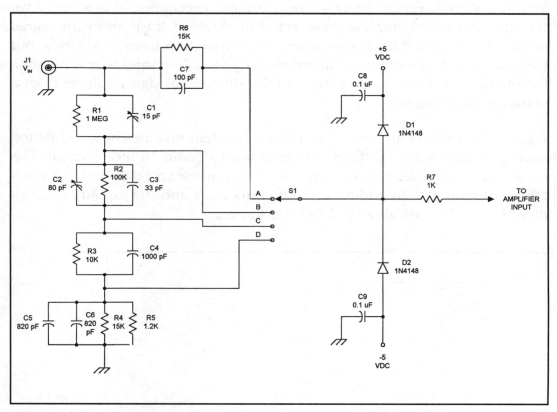

*Figure 16-3. Input step attenuator network.*

## Guard Shielding

One of the properties of the differential amplifier, including the instrumentation amplifier, is that it tends to suppress interfering signals from the environment. The common mode rejection process is at the root of this capability. When an amplifier is used in a situation where it is connected to an external signal source through wires, those wires are subjected to strong local 60 Hz ac fields from nearby power line wiring. Fortunately, in the case of the differential amplifier the field affects both input lines equally, so the induced interfering signal is cancelled out by the common mode rejection property of the amplifier.

Unfortunately, the cancellation of interfering signals is not total. There may be, for example, imbalances in the circuit that tend to deteriorate the CMRR of the amplifier. These imbalances may be either internal or external to the amplifier circuit. Figure 16-4a shows a common scenario. In this figure we see the differential amplifier connected to shielded leads from the signal source, $V_{in}$. Shielded lead wires offer some protection from local fields, but there is a problem with the standard wisdom regarding shields: it is possible for shielded cables to manufacture a valid differential signal voltage from a common mode signal!

Figure 16-4b shows an equivalent circuit that demonstrates how a shielded cable pair can create a differential signal from a common mode signal. The cable has capacitance between the center conductor and the shield conductor surrounding it. In addition, input connectors and the amplifier equipment internal wiring also exhibits capacitance.

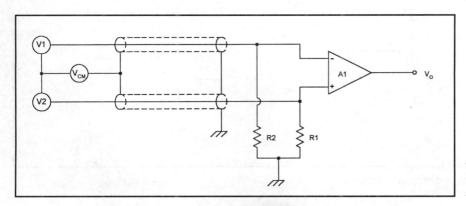

*Figure 16-4a. Shielded input leads circuit.*

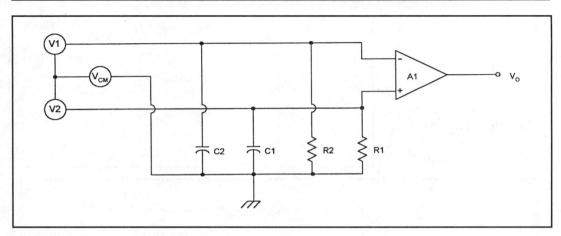

*Figure 16-4b. Equivalent circuit.*

These capacitances are lumped together in the model of Figure 16-4b as $C_{S1}$ and $C_{S2}$. As long as the source resistances and shunt resistances are equal, and the two capacitances are equal, there is no problem with circuit balance. But inequalities in any of these factors (which are commonplace) creates an unbalanced circuit in which common mode signal $V_{cm}$ can charge one capacitance more than the other. As a result, the difference between the capacitance voltages, $V_{CS1}$ and $V_{CS2}$, is seen as a valid differential signal.

A low-cost solution to the problem of shield-induced artifact signals is shown in Figure 16-5a. In this circuit a sample of the two input signals are fed back to the shield, which in this situation is not grounded. Alternatively, the amplifier output signal is used to drive the shield. This type of shield is called a guard shield. Either double shields (one on each input line) as shown, or a common shield for the two inputs can be used.

An improved guard shield example for the instrumentation amplifier is shown in Figure 16-5b. In this case a single shield covers both input lines, but it is possible to use separate shields. In this circuit a sample of the two input signals is taken from the junction of resistors R8 and R9, and fed to the input of a unity gain buffer/driver "guard amplifier" (A4). The output of A4 is used to drive the guard shield.

Perhaps the most common approach to guard shielding is the arrangement shown in Figure 16-5c. Here we see two shields used; the input cabling is double-shielded insulated wire. The guard amplifier drives the inner shield,

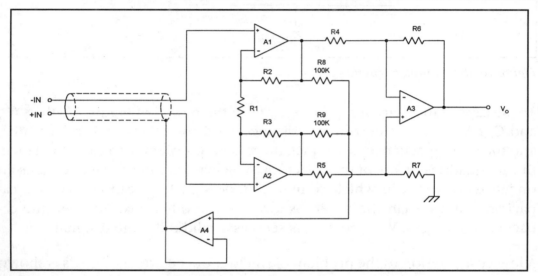

*Figure 16-5a. Guard shield circuit eliminates problems caused by the Figure 16-a circuit.*

*Figure 16-5b. Active guard shield uses an amplifier to drive the shield.*

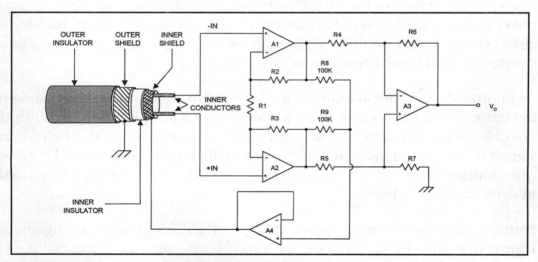

*Figure 16-5c. Double shielded version provides best protection.*

which serves as the guard shield for the system. The outer shield is grounded at the input end in the normal manner, and serves as an electromagnetic interference suppression shield.

**Guard Shielding on Printed Circuit Boards.** The guard shield concept can be extended to the input pins of high gain amplifiers on printed circuit boards. Figure 16-6 shows the method for placing a guard ring around a printed circuit pad (e.g. an IC pin). The ring is grounded in most cases, but is always connected to the guard shield of the input cable. The center conductor of the cable is connected to the pad itself. The version shown in Figure 16-6 keeps one end of the ring open in order to accommodate connection to the PC pad. If the board is two-sided or multi-layered, then the ring can be complete, with the connection to the pad occurring on the top or on one of the intermediate layers.

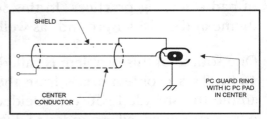

*Figure 16-6. Connecting guard shield to printed circuit board in a high gain application.*

**Guard Shielding Specialized Amplifiers.** There are a number of specialist integrated circuits on the market. These are often designed specially for the analog subsystem of PC-based instruments, although in many cases they are more general in format. An example of a general IC amplifier is the B-B INA-101AG device shown in Figure 16-7. It is an integrated circuit instrumentation amplifier (ICIA) device, with the gain being set by an external resistor RG. This type of device has provision for guard shield connections directly to the IC (pins 4 and 8). These pins are fed to the separate inputs of a summing amplifier, the output of which drives the guard shield.

## Grounding and Ground Loops

Impulse noise due to electrical arcs, lightning bolts, electrical motors and other devices can interfere with the operation of sensors and their associated circuits. Shielding of lines (see above) will help somewhat, but it isn't the entire answer. Filtering (discussed below) is useful, but it is at best a two-edged sword, and it must be done prudently and properly. Filtering on signal lines tends to broaden fast risetime pulses, and attenuate high frequency signals...and in some circuits causes as many problems as it solves.

Other electrical devices nearby can induce signals into the instrumentation system, the chief among these sources being the 60 Hz ac power system. It is wise to use only differential amplifier inputs, because of their high common mode rejection ratio. Signals from the desired source can be connected across the two inputs, and so become a differential signal, while the 60 Hz ac interference tends to affect both inputs equally (so is common mode).

It is sometimes possible, however, to manufacture a differential signal from a common mode signal. Earlier we said this phenomenon can occur because of bad shielding practices. In this section we are going to expand on that theme and consider grounds as well as shields.

One source of this problem is called a ground loop, and is shown in Figure 16-8a. This problem arises from the use of too many grounds. In this example the shielded source, shielded input lines, the amplifier and the dc power supply are all grounded to different points on the ground plane. Power supply dc currents (I) flow from the power supply at point A to the amplifier power common at point E. Since the ground has ohmic resistance, albeit very low resistance, the voltage drops E1 through E4 are formed. These voltages are seen by the amplifier as valid signals, and can become especially troublesome if I is a varying current.

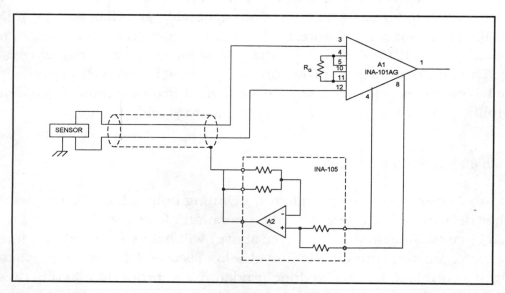

*Figure 16-7. INA-101AG instrumentation amplifier with active guard shield driver.*

The solution to this problem is to use single-point grounding, as shown in Figure 16-8b. Some amplifiers used in sensitive graphics or CRT oscilloscopes keep the power and signal grounds separate, except at some single, specific common point. In fact, a few models go even further by creating several signal grounds, especially where both analog and digital signals might be present.

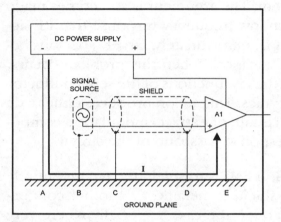

Figure 16-8a. Standard grounding causes ground loops due to IR voltage drops.

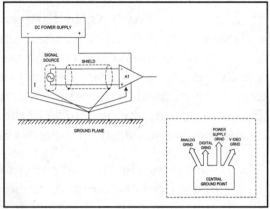

Figure 16-8b. Single-point grounding eliminates ground loops.

In some instances the shield on the input lines must not be grounded at both ends. In those cases, it is usually better to ground only the amplifier ends of the cables.

## Electromagnetic Interference (EMI) to Sensor Circuits

Most sensors and associated circuits are susceptible to electromagnetic interference (EMI) because of the typically very low signal levels produced by those sensors. It is worthwhile, therefore, to review the causes of, and possible cures for, these types of EMI problem. To identify fixes for this type of EMI case it is helpful to know the typical routes that EMI signals follow to enter a sensor system.

## EMI Transmission Paths

There are three principal transmission paths for EMI into a sensor system: penetration, leakage, and conduction. Figure 16-9a illustrates penetration. In this case, the impinging electromagnetic (EM) wave cuts across the sensor and its circuit, setting up RF currents in the sensor circuitry. These RF currents become valid signals in the system. The argument is sometimes made that sensor circuits typically have such low frequency response that RF signals won't affect them. This argument is, unfortunately, largely specious because of phenomena such as autorectification. When this problem occurs, the RF signals rectify (e.g. "detect") in the PN junctions of the semiconductor components used in the sensor electronics. This action produces either a dc bias that distorts the operation of the circuit, or extracts modulation components that are within the frequency response passband of the circuit.

The solution for eliminating penetration EMI is to shield the circuit entirely in a shielded environment. A proper shield is metallic, or another conductive material, and is well sealed. It is often the case, especially where very high frequencies or microwaves are causing the interference, that the shielded box holding the electronics must have additional shielding added. In many cases, extra fasteners or screws are needed to seal the lips or edges of the shielded cabinet. The rule of thumb is that the screws should be spaced less than one-half wavelength at the highest frequency expected (or specified for that particular product).

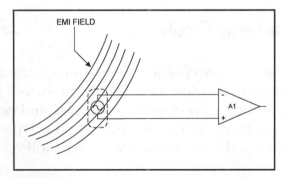

*Figure 16-9a. EMI field affecting an unprotected sensor circuit.*

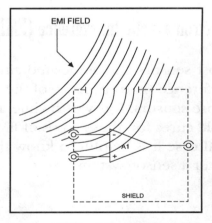

*Figure 16-9b. Effect of shielding in open box.*

Leakage occurs when the EM wave enters the shielded cabinet through small cracks or spaces in the shielding (Figure 16-9b). The standard wisdom is that a half wavelength slot will admit RF to a shielded compartment. Unfortunately, that is not always true, especially where non-sinusoidal RF waveforms are present. The wavelength of the signal fundamental frequency is:

$$\lambda_{meters} = \frac{300}{F_{mhz}} \qquad eq.\,(16\text{-}2)$$

Thus, a 150 MHz taxicab transmitter creates a 2 meter wavelength signal, so is not a problem in most cases because the necessary gap (1 meter) is wide compared with the size of the instrument cabinet. In the microwave region, however, wavelengths drop to the centimeter and millimeter region, so these become a larger problem.

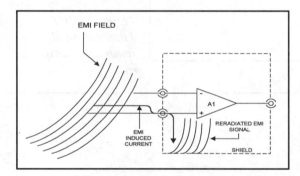

Figure 16-9c.
Effect in shielded box
showing reradiation.

Even at lower frequencies, however, modulation or other situations can raise the number and strength of the harmonics of the fundamental, and these can cause severe interference under the right circumstances. Pulsed signals, for example, can have significant harmonics of 100 times the fundamental frequency. A 400 MHz signal has a wavelength of 0.75 meters. If it is a sine wave signal, or contains linear voice modulation (or certain other forms), then there will be few harmonics. But if the 400 MHz signal is pulsed, then it might easily have harmonics to 4 GHz or even 40 GHz. At these frequencies, the shielding must be tight.

There are two modes of leakage. The resonant mode occurs when the gap in the sensor shielding is resonant at the interfering frequency. If the gap is a half wavelength, then it will act as a slot antenna and efficiently reradiate the offending signal inside the shielded enclosure.

Conduction EMI (Figure 16-9c) occurs when a signal induces current into an exposed power, ground, control or signal cable entering or leaving the shielded compartment. In this type of interference, the EMI signal reradiates inside the housing, and is picked up by the sensor or its electronics. Surprisingly short lengths of wire will act as relatively efficient antennas in the UHF region and above.

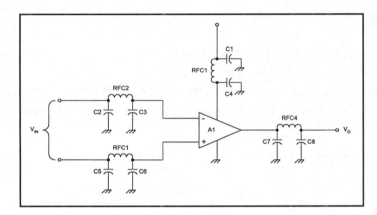

*Figure 16-10a.*
*Filtering on all*
*leads of an amplifier.*

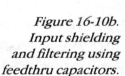

*Figure 16-10b.*
*Input shielding*
*and filtering using*
*feedthru capacitors.*

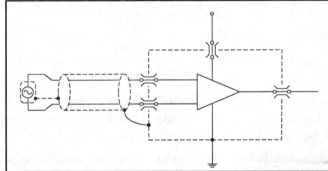

## Some Cures

Figure 16-10 shows two common fixes for EMI in sensor systems: filtering (Figure 16-10a) and shielding (Figure 16-10b). It is likely that both will have to be used in any given system.

The use of filtering to attenuate RF signals before they get to the sensor or electronics circuits is shown in Figure 16-10a. Each line in the system (whether power or signal) has a p-section filter consisting of an inductor element called an rf choke (RFC), and two capacitors. An example in Figure 16-10a is RFC1/C1/C4 in the +POWER line. The p-section filter is basically a low-pass RF filter that has a -3 dB cut-off frequency that is well below the offending signal's frequency, but well above the highest frequency in the Fourier spectrum of the sensor output signal. A rule of thumb for the filter cut-off frequency is to calculate it from the length of exposed cable in the sensor system:

$$F_c \ll \frac{300}{2L} \qquad\qquad \text{eq. (16-3)}$$

Where:

Fc is the -3 dB cut-off frequency of the low-pass filter.

L is the length of cable in meters.

For example a 1 foot (0.295 meter) cable is half wavelength resonant at a frequency of 509 MHz. Signal interference is maximum at that frequency, but is still significant at frequencies removed from Fc, especially those that are higher. The "much less than" (<<) symbol in the equation above can be taken to mean 1/10, so the cut-off frequency, $F_c$, of the low-pass filter should be $F_c/10$, or in the case above 509 MHz/10 = 50.9 MHz.

The filter should be installed as close to the connector in the shielded bulkhead as possible. Otherwise, reradiation can occur from the wiring between the connector and the filter. Commercially available EMI filters usually are chassis or "bulkhead" mounted, so this requirement is easily met most of the time in actual practice.

If the EMI filter is installed in an ac power line, then make absolutely certain that it is of a type that is specifically designed for such service (the voltage rating is not always a sufficient indicator - look for the words "ac line" in the specification). If a filter is not specifically designed for service in 110 Vac or 220 Vac lines, then using them in that service can be dangerous.

Double shielding combined with filtering is shown in Figure 16-10c. The sensor is located inside of an inner shielded compartment that is, itself, enclosed within the outer shielded compartment. An EMI filter (RFC1/C1/C2) is placed between the cable strain relief and the inner shield compartment.

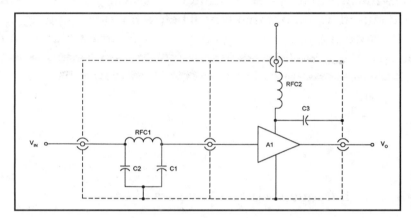

*Figure 16-10c.*
*Use of a low-pass*
*filter on input.*

Figure 16-11 shows the same approach for a differential input amplifier. In this case, the L and C elements are placed in both lines of the differential amplifier. In some cases, resistors are also placed in series with the amplifier inputs, either in lieu of the inductors or in concert with them. The goal is to make the resistance low compared with the input impedance of the amplifier, and high compared with the reactance of the capacitors at the cut-off frequency. In that case, the resistor-capacitor elements serve as a low-pass filter.

*Figure 16-11.*
*Multiple RCL*
*input filters.*

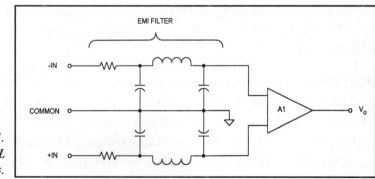

## Conclusion

Artifacts in analog signal data can be largely eliminated by the proper use of the techniques discussed in this chapter. While the topic is worthy of a book in its own right (and indeed, several have been written), these techniques are suitable for many - perhaps most - practical applications.

# Chapter 17

# A/D and D/A Converters

Data converters do one of two jobs, either: 1) convert a binary digital word to an equivalent analog current or voltage, or, 2) convert an analog current or voltage to an equivalent binary word. The former are called digital-to-analog converters (DACs), while the latter are called analog-to-digital converters (A/D or ADC). These devices form the interface between digital computers and the analog world, which include most sensors and actuators.

## Approaches to Data Conversion

There are several approaches to converting data from electronic circuits and scientific instruments into digital binary numbers required by a digital computer. This job requires an A/D converter. The opposite type of converter (DAC) does the opposite work. It converts the binary output from a computer into an analog voltage or current. In discussing data converters it is best to begin with a discussion of DACs because many A/D circuits contain embedded DACs.

The difference between A/D and DAC functions are shown in Figures 17-1a and 17-1b. In Figure 17-1a, an A/D converter is used to translate an analog voltage waveform into the equivalent binary numbers that the computer can

use. The waveform is simplified compared with actual waveforms, but that is
done to make it easier for me to write and you to read. The waveform achieves
certain values of voltage between 0 and 4 volts, as follows:

| Time | Voltage |
|------|---------|
| T1 | 0 volts |
| T2 | 1 volt |
| T3 | 2 volts |
| T4 | 3 volts |
| T5 | 4 volts |
| T6 | 3 volts |
| T7 | 2 volts |
| T8 | 1 volt |
| T9 | 0 volts |

This voltage waveform is applied to the analog input of the A/D converter,
which produces the following output binary numbers at the times indicated:

| Time | Binary Output | Decimal Equivalent |
|------|---------------|--------------------|
| T1 | $0000_2$ | 0 volts |
| T2 | $0001_2$ | 1 volt |
| T3 | $0010_2$ | 2 volts |
| T4 | $0011_2$ | 3 volts |
| T5 | $0100_2$ | 4 volts |
| T6 | $0011_2$ | 3 volts |
| T7 | $0010_2$ | 2 volts |
| T8 | $0001_2$ | 1 volts |
| T9 | $0000_2$ | 0 volts |

In other words, the A/D converter takes the analog input voltage, converts to
a series of binary (base-2) numbers ("words") that represent the values, and
then forwards these binary numbers to an input port of the computer.

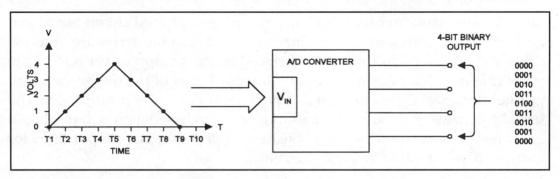

*Figure 17-1a. The Analog-to-Digital conversion function.*

The DAC process is just the opposite (Figure 17-1b). In this case, the digital computer generates a series of binary values that represent a data pattern from which the waveform could be constructed. The DAC will convert these binary numbers into equivalent voltages (or currents in many models), and output them to instruments such as voltmeters, ammeters, oscilloscopes or mechanical (paper) chart recorders.

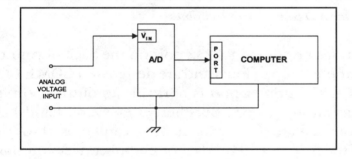

*Figure 17-1b. The Digital-to-Analog conversion function.*

Note the difference in waveforms between Figures 17-1a and 17-1b. They are supposedly the same because the same set of binary numbers is associated with each. The waveform in Figure 17-1b is what you get if you digitize the waveform in Figure 17-1a, and then feed it to a DAC for reconstruction of the analog waveform. The distortion is due to the fact that too few bits were used to represent the waveform. If more bits (e.g. 12-bits or 16-bits) were used, then the waveform in Figure 17-1b would more nearly resemble that of Figure 17-1a.

The A/D converter can be internal, i.e. on a plug-in printed circuit board that is interfaced directly with the internal computer data bus structure. Alternatively, the A/D converter can be interfaced to the computer via one of the ports (Figure 17-2). There are three different forms of port: 8-bit or 16-bit parallel port, serial communications port and Centronics printer port. The straight parallel ports are rare, and must be purchased as separate plug-in cards for most PCs. These ports consist of 8 or 16 separate input lines for carrying data (one line for each data bit).

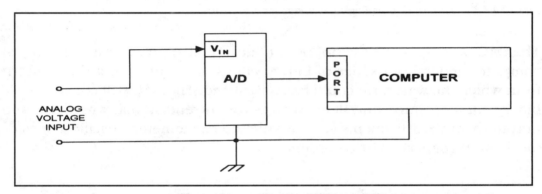

*Figure 17-2. Interfacing the A/D converter to computer.*

The serial communications port is the RS-232 port on the back of your computer. It's on most computers (up to four) and are designated COM1, COM2, COM3 or COM4 on PCs. The printer port is normally an output port, so is not often thought of as an input port. But there are several handshaking lines on the printer port that are bidirectional. They can be used to trigger the START function of the A/D, or receive data or end-of-conversion signals.

Many of the A/D converters that we will examine are based on the DAC circuit, so (even though DACs per se are of less interest to most) we will look at DAC circuitry first.

## Digital-to-Analog Converters (DACS)

There are several different approaches to DAC design, but all of them are varieties of a weighted current or voltage system that accepts binary words by appropriate switch contacts. The most common example is the R-2R ladder shown in Figure 17-3. The active element, A1, is an operational amplifier in a unity gain inverting follower configuration. In the circuit of Figure 17-3 the digital inputs are shown as mechanical switches, but in a real data converter circuit the switches would be replaced by electronic switching devices (e.g. transistors). The electronic switches are driven by either a binary counter, or an N-bit parallel data line.

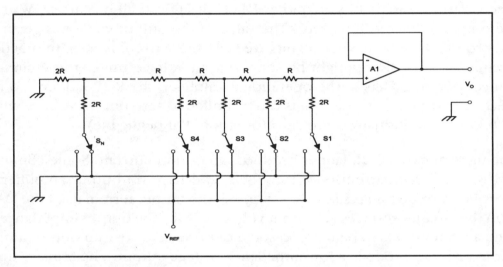

*Figure 17-3.  R-2R binary ladder.*

A precision reference voltage source ($V_{ref}$) is required for accurate data conversion. This voltage is most often +2.56 volts, +5.00 volts or +10.00 volts. Other voltages can also be used, however. The accuracy of the converter is dependent upon the precision of the reference voltage source. There are other sources of error, but if the reference voltage accuracy is poor, then there is no hope for any other factors to be effective in improving the performance of the circuit. I can recall situations where expensive 16-bit data converters were used with zener diode reference sources that had so much thermal drift that they were suited only to 4 to 6-bit models. Although almost any voltage regulator can be pressed into service as the reference, it is prudent to select a precision, low-drift model. Consider the following situation: a 12-bit A/D converter breaks up the input voltage range into 4,096 different discrete values, or if one is zero 4,095 different non-zero values. Each step, produced by a change in the least significant bit of the 12-bit binary word, represents 0.0024 volts, or 2.4 µV (called the "1-LSB" value). On a "0 to 10 volt" A/D converter, therefore, the maximum input voltage is 9.9976 volts (with some doubt in the last position). If a zener diode with a thermal drift of ±0.100 volts (100 µV) is used as the reference, then the uncertainty arises at the 9.90 volt level! This is a good way to make a 12-bit A/D converter act like a 6 or 7-bit A/D!

Returning to Figure 17-3, consider the circuit action under circumstances where various binary bits are either HIGH or LOW. If all bits are LOW, then the output voltage will be zero. The value of the output voltage is given by the product I × R, and when all bits are LOW this current is zero. In practical circuits, though, there might be some output voltage under these circumstances due to offsets in the operational amplifier, the R-2R ladder and the electronic switches. These offsets can be nulled to zero output voltage when all bits are intentionally set to zero (or ignored, if negligible).

The unterminated R-2R ladder produces an output current. Some commercial IC DACs are current output models, and have no output amplifier. If there is a terminating resistor $(R_t)$ shunting the output terminals of the DAC, then the circuit produces an output voltage $I_o \times R_t$. The output impedance of such a circuit tends to be high, so some of these DACs use an output amplifier in order to produce a low-impedance voltage output. The transfer function of the R-2R ladder type of DAC is:

$$V_o = \frac{V_{ref} A}{2^N} \qquad\qquad eq.\ (17\text{-}1)$$

Where:

   $V_o$ is the output potential.

   $V_{ref}$ is the reference potential.

   A is the decimal value of the applied binary word.

   N is the number of bits in the applied binary word.

If the most significant bit (MSB) is made "1" (i.e. HIGH), then the output voltage will be approximately $V_{ref}/2$. Similarly, if the next most significant bit is turned on (set to HIGH) and all others are LOW, then the output will be $V_{ref}/4$. The least significant bit (LSB) would contribute $V_{ref}/2N$ to the total output voltage. For example, with an 8-bit DAC, the LSB changes the output

$V_{ref}/28$, or $V_{ref}/256$. This change is called the 1-LSB value because it is the change that occurs in response to a change in the least significant bit. Figure 17-4 graphs the output of a voltage DAC in response to the entire range of binary numbers applied to the digital inputs. The result is a staircase waveform that rises by the 1-LSB value for each 1-LSB change of the binary word. This step height represents the minimum discernible resolution of the circuit.

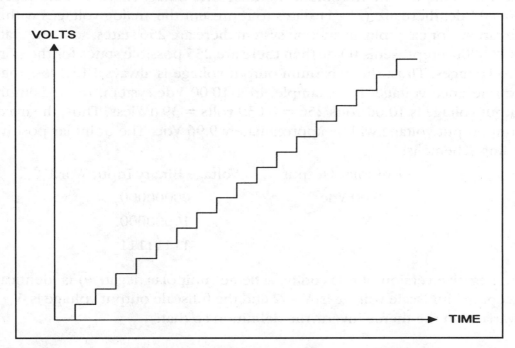

*Figure 17-4. Waveform of a DAC output when input runs through the binary numbers.*

The reference source can be either internal or external to an integrated circuit DAC. If the reference voltage or current source is external, then the DAC is said to be a multiplying DAC or MDAC. The multiplication takes place between the analog reference and the fraction defined as $A/2^N$ in Equation 17-1. If the reference source is completely internal, and not adjustable (except for fine trimming), then the DAC is said to be a non-multiplying DAC or simply "DAC."

## Coding Schemes

There are several different coding methods for defining the transfer function of the DAC. The most common of these are: unipolar positive, unipolar negative, symmetrical bipolar and asymmetrical bipolar.

The unipolar coding schemes provide an output voltage of one polarity only. These circuits usually produce 0 volts for the minimum and some positive or negative value for the maximum. Because one binary number state represents 0 Vdc, there are ($2^N$ - 1) states to represent the analog voltages within the range. For example, in an 8-bit system there are 256 states, so if one state ($00000000_2$) represents 0 Vdc then there are 255 possible states for the non-zero voltages. Thus, the maximum output voltage is always 1-LSB less than the reference voltage. For example, in a 10.00 Vdc system, the maximum output voltage is 10.00 volts/256 = 0.039 volts = 39 μV less. Thus, the maximum output voltage will be approximately 9.96 Vdc. The unipolar positive coding scheme is:

| Unipolar Output | Voltage Binary Input Word |
|---|---|
| 0.00 Vdc | $00000000_2$ |
| $V_{max}/2$ | $10000000_2$ |
| $V_{max}$ | $11111111_2$ |

The negative version of this coding scheme (unipolar negative) is identical, except the midscale voltage is $-V_{max}/2$ and the fullscale output voltage is $-V_{max}$. A variant on the theme inverts the definition so that:

| Unipolar Output | Voltage Binary Input Word |
|---|---|
| 0.00 Vdc | $11111111_2$ |
| $V_{max}/2$ | $1000000_2$ |
| $V_{max}$ | $00000000_{@-\{2\}}$ |

The bipolar coding scheme faces a difficulty that requires a trade-off in the design. There is an even number of output states in a binary system. For example, in the standard 8-bit system there are 256 different output states. If one state is selected to represent 0 Vdc, then there are 255 states left to represent the voltage range. As a result, there is an even number of states to represent positive and negative states either side of 0 Vdc. For example, 127 states might be assigned to represent negative voltages, and 128 to represent positive voltages. In the asymmetrical bipolar coding, therefore, the pattern might look like:

| Bipolar Output Voltage | Binary Input Word |
|---|---|
| $-V_{max}$ | $00000000_2$ |
| 0.00 Vdc | $10000000_2$ |
| $+V_{max}$ | $11111111_2$ |

A decision must be made regarding which polarity will lose a small amount of dynamic range.

The other bipolar coding system is the symmetrical bipolar scheme. The decision in the symmetrical scheme is that each polarity will be represented by the same number of binary states either side of 0 Vdc. But this scheme does not permit a dedicated state for zero. The scheme is:

| Bipolar Output Voltage | Binary Input Word |
|---|---|
| $-V_{max}$ | $00000000_2$ |
| -Zero (-1 LSB) | $01111111_2$ |
| 0.00 Vdc | (disallowed) |
| +Zero (+1 LSB) | $10000000_2$ |
| $+V_{max}$ | $11111111_2$ |

The state "plus zero" is more positive the 1-LSB value than 0 Vdc, while the "minus zero" state is more negative than 0 Vdc by the same 1-LSB value.

# A Practical IC DAC Example

A number of different manufacturers offer low-cost IC DACs that contain almost all of the circuitry needed for the process, except possibly the reference source (although some devices do contain the reference source also) and some operational amplifiers for either level shifting or current-to-voltage conversion.

For purposes of this chapter the DAC-08 device is used as a practical circuit example. This eight-bit DAC is now something of an industry standard, and is available from several sources. This DAC is sometimes designated LMDAC-0800. An easily available, and closely related, device is the DAC-0806.

Figure 17-5a shows the basic circuit configuration for the DAC-08. In subsequent circuits, the power supply terminals are deleted for simplicity's sake; they will always be the same as shown here. The internal circuitry of the DAC-08 is the R-2R ladder shown in the previous section, but has two outputs: $I_o$ and NOT-$I_o$. These current outputs are unipolar and complementary (Figure 17-5b); if the full-scale output current is $I_{max}$, then $I_{max} = [I_o + $ NOT-$I_o]$. The specified value of $I_{max}$ on the DAC-08 family of DACs is 2 mA.

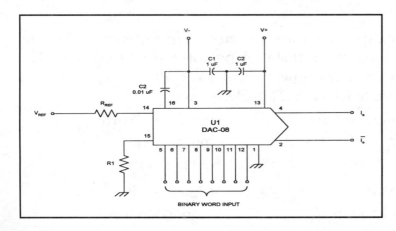

Figure 17-5a.
Basic DAC-08 circuit.

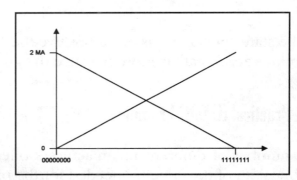

Figure 17-5b.
Output function.

Two types of input signal are required to make this DAC work: an analog reference and an 8-bit digital signal. The analog signal the reference current, $I_{ref}$, is applied through pin no. 14. This current may be generated by combining a precision reference voltage source with a precision, low temperature coefficient, resistor to convert $V_{ref}$ to $I_{ref}$. Alternatively, a constant current source may be used to provide Iref. For TTL compatibility of the binary inputs, make $V_{ref}$ = 10.000 volts, and $R_{ref}$ = 5000 ohms.

The other type of input is the 8-bit digital word, which is applied to the IC at pins 5 through 12, as shown. The logic levels which operate these inputs can be preset by the voltage applied to pin no. 1 (for TTL operation, pin no. 1 is grounded). In the TTL-compatible configuration shown, LOW is 0 to 0.8 volts, while HIGH is +2.4 to +5 volts.

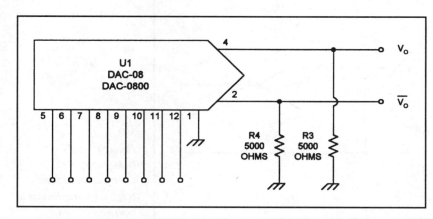

*Figure 17-6. 0-10 volt, TTL compatible circuit with voltage output.*

Figure 17-6 shows the connection of the DAC-08 (less power supply and reference input) required to provide the simplest form of unipolar operation over the range of approximately 0 to 10 volts. When the input word is $00000000_2$, then the DAC output is 0 volts, plus-or-minus the dc offset error. A half-scale voltage (-5 volts) is given when the input word is $10000000_2$. This situation occurs when the MSB is HIGH and all other digital inputs are LOW. The full-scale output will exist only when the input word is $11111111_2$ (all HIGH). The output under full-scale conditions will be -9.96 volts, rather than -10 volts as might be expected (note: -9.96 volts is 1-LSB less than -10 volts). These codes are:

| Condition | Binary Code | $V_o$ |
|---|---|---|
| Full-scale | 11111111 | -9.96 V |
| Half-scale | 10000000 | -5.00 V |
| Zero | 00000000 | 0.0 V |

The circuit in Figure 17-6 works by using resistors R3 and R4 as current-to-voltage converters. When currents $I_o$ and NOT-$I_o$ pass through these resistors, a voltage drop of IR, or $5.00 \times I_o$ (mA) is created. A problem with this circuit is that is has a high source impedance (5 kohms, with the values shown for R3/R4).

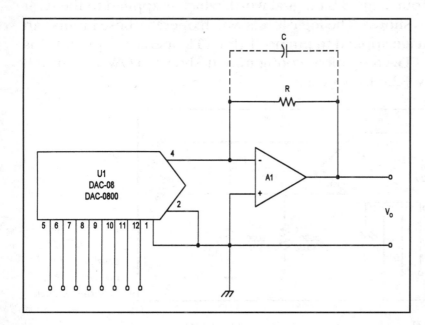

Figure 17-7.
0-10 volt, TTL
compatible circuit
with low-impedance
voltage output.

Figure 17-7 shows a simple method for converting Io to an output voltage ($V_o$) with a low output impedance (less than 100 ohms) by using an inverting follower operational amplifier. The output voltage is simply the product of the output current and the negative feedback resistor:

$$V_o = R \times I_o \qquad\qquad eq. (17\text{-}2)$$

As in the case previously described, a 5000 ohm resistor and a 10.00 Vdc reference voltage will produce a -9.96 volt output voltage when the DAC-08 is set up for TTL inputs and 2.0 mA $I_{o(max)}$.

The frequency response of the DAC circuit can be tailored to meet certain requirements. The normal output waveform of the DAC is a staircase when the digital input increments up from $00000000_2$ to $11111111_2$ in a monotonic manner. In order to make the staircase into an actual ramp function, a low-pass filter is needed at the output to remove the "stepness" of the normal waveform. A capacitor shunted across the feedback resistor (R) will offer limited filtering on the order of -6 dB/octave above a cut-off frequency of:

$$\bullet \quad F = \frac{1,000,000}{2 \pi R C_{\mu F}} \qquad \qquad eq.\ (17\text{-}3)$$

Where:

F is the -3 dB frequency in hertz (Hz).

R is in ohms ($\Omega$).

$C_{\mu F}$ is in microfarads ($\mu F$).

In most practical circuits the required value of F is known from the application. It is the highest frequency Fourier component in the input waveform. It is necessary to calculate the value of capacitor needed to achieve that cut-off frequency, so would swap the F and C terms:

$$\bullet \quad C_{\mu F} = \frac{1,000,000}{2 \pi R F} \qquad \qquad eq.\ (17\text{-}4)$$

## A/D Converters

The analog-to-digital converter (A/D) is used to convert an analog voltage or current input to an output binary word that can be used by a computer. Of the many techniques that have been published for performing an A/D conversion, only a few are of interest to us; so we will consider only the voltage-to-frequency, single-slope integrator, dual-slope integrator, counter (or servo), successive approximation and flash methods. The basic size of circuit that we will show is the 8-bit A/D converter, which for many purposes is all that is needed. These same discussions are also useful for 10-bit, 12-bit, or higher order A/D converters.

## Integration A/D Methods

Most digital panel meters (DPM) and digital multimeters (DMM) use either the single integration or dual-slope integration methods for the A/D conversion process. An example of a single-slope integrator A/D converter is shown in Figure 17-8a, while its timing diagram is shown in Figure 17-8b. The single-slope integrator is simple, but is limited to those applications that can tolerate accuracy of one or two percent.

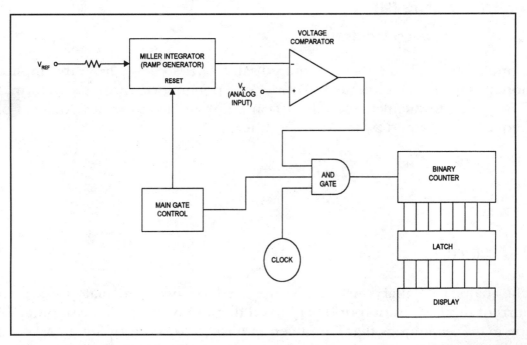

*Figure 17-8a. Single-slope A/D converter.*

The single-slope integrator A/D converter of Figure 17-8a consists of five basic sections: ramp generator, comparator, logic, clock and an output encoder consisting of a binary counter, latch and display. The ramp generator is an ordinary operational amplifier Miller integrator with its input connected to a stable, fixed, reference voltage source. This makes the input current $I_{ref}$ essentially constant; so the voltage at point-B will rise in a nearly linear manner, creating the voltage ramp.

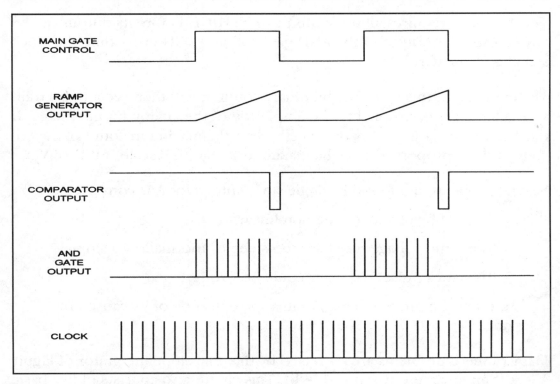

*Figure 17-8b. Timing diagram.*

The comparator is an operational amplifier that has an open feedback loop. The circuit gain is the open-loop gain ($A_{vol}$) of the device selected - typically very high even in low-cost operational amplifiers. When the analog input voltage $V_x$ is greater than the ramp voltage, the output of the comparator is saturated at a logic-HIGH level.

The logic section consists of a main AND gate, a main-gate control, and a clock. The waveforms associated with this circuit are shown in Figure 17-8b. When the output of the main-gate control logic section is LOW, switch S1 remains closed, so the ramp voltage is zero. The main-gate signal at point-A is a low frequency square wave with a frequency equal to the desired time-sampling rate. When point-A is HIGH, S1 is open, so the ramp will begin to rise linearly. When the ramp voltage is equal to the unknown input voltage $V_x$, the differential voltage seen by the comparator is zero; so its output drops LOW.

The AND gate requires all three inputs to be HIGH before its output can be HIGH also. The output of the AND gate will go HIGH every time the clock signal is also HIGH.

The encoder, in this case an 8-bit binary counter, will then see a pulse train with a length proportional to the amplitude of the analog input voltage. If the A/D converter is designed correctly, then the maximum count of the encoder will be proportional to the maximum range (full-scale) value of $V_x$.

Several problems are found in single-slope integrator A/D converters:

1. The ramp voltage may be non-linear.

2. The ramp voltage may have too steep or too shallow a slope.

3. The clock pulse frequency could be wrong.

4. It may be prone to changes in apparent value of Vx caused by noise.

Many of these problems are corrected by the dual-slope integrator of Figutr 17-9a (timing circuit in Figure 17-9b). This circuit also consists of five basic sections: integrator, comparator, control logic section, binary counter, and a reference current or voltage source. An integrator is made with an operational amplifier connected with a capacitor in the negative feedback loop, as was the case in the single-slope version. The comparator in this circuit is also the same sort of circuit as was used in the previous example. In this case, though, the comparator is ground-referenced by connecting +IN to ground.

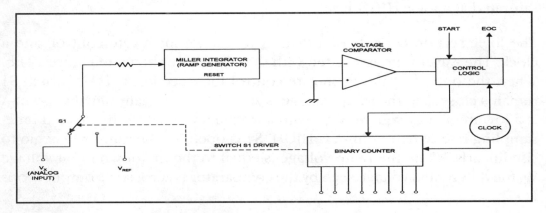

*Figure 17-9a. Dual-slope A/D converter.*

When a start command is received, the control circuit resets the counter to 00000000$_2$, resets the integrator to 0-volts (by discharging C1 through switch S1), and sets electronic switch S2 to the analog input (position-A). The analog voltage creates an input current to the integrator which causes the integrator output to begin charging capacitor C1; the output voltage of the integrator will begin to rise. As soon as this voltage rises a few millivolts above ground potential (0 Vdc) the comparator output snaps HIGH-positive. A HIGH comparator output causes the control circuit to enable the counter, which begins to count pulses.

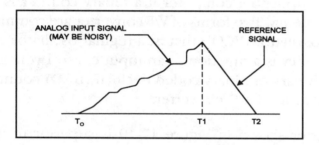

*Figure 17-9b.*
*Timing diagram.*

The counter is allowed to overflow, and this output bit sets switch S2 to the reference source (position-B). The graph of Figure 17-9b shows the integrator charging during the interval between start and the overflow of the binary counter. At time T2 the switch changes the integrator input from the analog signal to a precision reference source. Meanwhile, at time T2 the counter had overflowed, and again it has an output of 00000000$_2$ (maximum counter + 1 more count is the same as the initial condition). It will, however, continue to increment so long as we have a HIGH comparator output. The charge accumulated on capacitor C1 during the first time interval is proportional to the average value of the analog signal that existed between T1 and T2.

Capacitor C1 is discharged during the next time interval (T2-T3). When C1 is fully discharged the comparator will see a ground condition at its active input, so will change state and make its output LOW. Even though this causes the control logic to stop the binary counter, it does not reset the binary counter. The binary word at the counter output at the instant it is stopped is proportional to the average value of the analog waveform over the interval

(T1-T2). An end-of-conversion (EOC) signal is generated to notify the computer so that it knows the output data is both stable and valid (therefore ready for use).

## Voltage-to-Frequency Converters

These circuits are not A/D converters in the strictest sense, but are very good for representing analog data in a form that can be tape recorded on a low-cost audio machine, or transmitted over radio. The V/F converter output can also be used for direct input to a computer if a binary counter is used to measure the output frequency. Two forms of V/F converter are common. One is a voltage controlled oscillator (VCO); that is, a regular oscillator circuit in which the output frequency is a function of an input control voltage. If the VCO is connected to a binary or binary coded decimal (BCD) counter, then the VCO becomes a V/F form of A/D converter.

The type of V/F converter shown in Figure 17-10 is superior to the VCO method. The circuit is shown in Figure 17-10a, while the timing waveforms are shown in Figure 17-10b. The operation of this circuit is dependent upon the charging of a capacitor, although not an RC network as in the case of some other oscillator or timer circuits. The input voltage signal ($V_x$) is amplified (if necessary) by A1, and then converted to a proportional current level in a voltage-to-current converter stage. If the voltage applied to the input remains constant, so will the current output of the V-to-I converter (I).

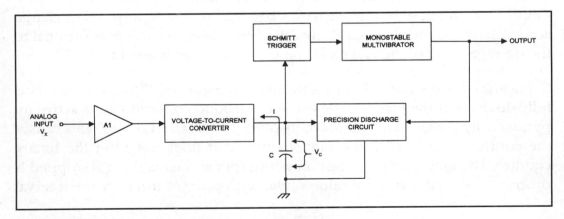

*Figure 17-10a. Voltage-to-Frequency converter.*

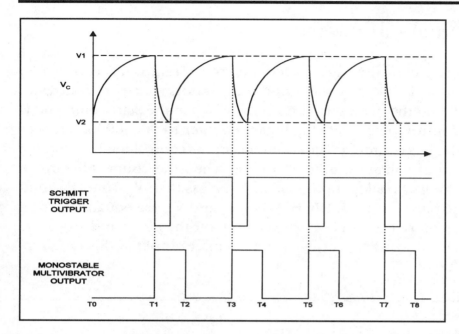

*Figure 17-10b.*
*Timing diagram.*

The current from the V-to-I converter is used to charge the timing capacitor, C. The voltage appearing across this capacitor ($V_c$) varies with time as the capacitor charges (see the $V_c$ waveform in Figure 17-10b). The precision discharge circuit is designed to discharge capacitor C to a certain level (V2) whenever the voltage across the capacitor reaches a predetermined value (V1). When the voltage across the capacitor reaches V1, a Schmitt trigger circuit is fired that turns on the precision discharge circuit. The precision discharge circuit, in its turn, will cause the capacitor to discharge rapidly but in a controlled manner to value V1. The output pulse snaps HIGH when the Schmitt trigger fires (i.e. the instant $V_c$ reaches V1) and drops LOW again when the value of $V_c$ has discharged to V2. The result is a train of output pulses whose repetition rate is exactly dependent upon the capacitor charging current, which, in turn, is dependent upon the applied voltage. Hence, the circuit is a voltage-to-frequency converter.

Like the VCO circuit, the output of the V/F converter can be applied to the input of a binary counter. The parallel binary outputs become the data lines to the computer. Alternatively, if the frequency is relatively low the computer can be programmed to measure the period between pulses. Also, certain computer interface chips have built-in timers that can measure the period.

## Counter Type (Servo) A/D Converters

A counter type A/D converter (also called "Servo" or "ramp" A/D converters) is shown in Figure 17-11. It consists of a comparator, voltage output DAC, binary counter, and the necessary control logic. When the start command is received, the control logic resets the binary counter to $00000000_2$, enables the clock, and begins counting. The counter outputs control the DAC inputs; so the DAC output voltage will begin to rise when the counter begins to increment. As long as analog input voltage $V_x$ is less than $V_{ref}$ (the DAC output), the comparator output is HIGH. When $V_x$ and $V_{ref}$ are equal, however, the comparator output goes LOW, which turns off the clock and stops the counter. The digital word appearing on the counter output at this time represents the value of $V_x$.

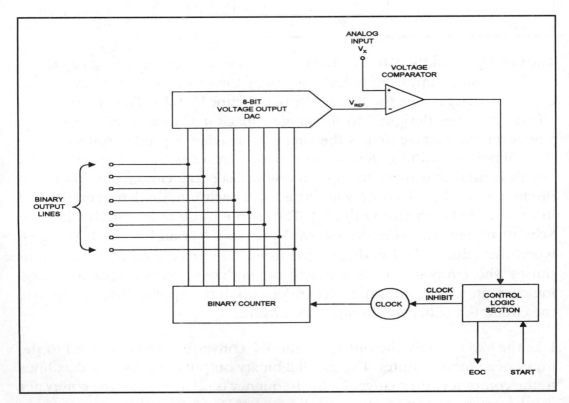

*Figure 17-11. Binary or "servo" A/D converter.*

Both slope and counter type A/D converters take too long for many applications, on the order of $2^N$ clock cycles (where N = number of bits). Conversion time becomes critical if a high frequency component of the input waveform is to be faithfully reproduced. Nyquist's criteria requires that the sampling rate (i.e. conversions per second) be at least twice the highest Fourier frequency to be recognized.

## Successive Approximation A/D Converters

Successive approximation A/D conversion is best suited for many applications where speed is important. This type of A/D converter requires only N+1 clock cycles to make the conversion, and some designs allow truncation of the conversion process after fewer cycles if the final value is found prior to N+1 cycles.

The successive approximation converter operates by making several successive trials at comparing the analog input voltage with a reference generated by a DAC. An example is shown in Figure 17-12. This circuit consists of a comparator, control logic section, a digital shift register, output latches and a voltage output DAC.

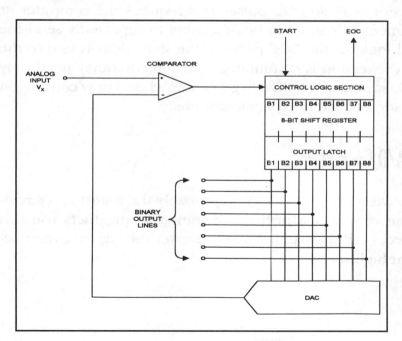

*Figure 17-12.*
*Successive*
*Approximation*
*Register A/D*
*converter.*

When a START command is received, a binary 1 (HIGH) is loaded into the MSB of the shift register, and this sets the output of the MSB latch HIGH. A HIGH in the MSB of a DAC will set the output voltage $V_{ref}$ to half-scale. If the input voltage $V_x$ is greater than $V_{ref}$, the comparator output stays HIGH and the HIGH in the shift register MSB position shifts one-bit to the right and therefore occupies the next most significant bit (bit 2). Again the comparator compares $V_x$ with $V_{ref}$. If the reference voltage from the DAC is still less than the analog input voltage, the process will be repeated with successively less significant bits until either a voltage is found that is equal to $V_x$ (in which case the comparator output drops LOW) or the shift register overflows.

If, on the other hand, the first trial with the MSB indicates that $V_x$ is less than the half-scale value of $V_{ref}$, the circuit continues making trials below $V_{ref}$. The MSB latch is reset to LOW and the HIGH in the MSB shift register position shifts one-bit to the right to the next most significant bit (bit-2). Here the trial is repeated again. This process will continue as before until either the correct level is found, or overflow occurs. At the end of the last trial (bit-8 in this case) the shift register overflows and the overflow bit becomes an EOC flag to tell the rest of the world that the conversion is completed.

This type, and most other types of A/D converters, require a starting pulse and signals completion with an EOC pulse. This requires the computer or other digital instrument to engage in bookkeeping to repeatedly send the start command and look for the EOC pulse. If the start input is tied to the EOC output, then conversion is continuous, and the computer need only look for the periodic raising of the EOC flag to know when a new conversion is ready. Such operation is said to be asynchronous.

## Parallel or "Flash" A/D Converters

The parallel A/D converter (Figure 17-13) is probably the fastest A/D circuit known. Indeed, the very fastest ordinary commercial products use this method. Some sources call the parallel A/D converter the "flash" circuit because of its inherent high speed.

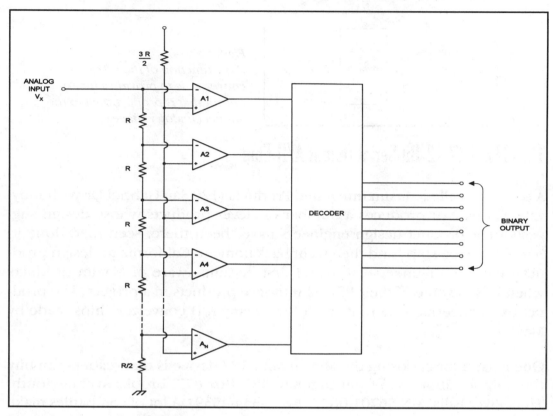

*Figure 17-13. Flash or parallel A/D converter.*

The parallel A/D converter consists of a bank of $(2^N - 1)$ voltage comparators biased by reference potential $V_{ref}$ through a resistor network that keeps the individual comparators 1-LSB apart. Since the input voltage is applied to all the comparators simultaneously, the speed of conversion is limited essentially by the slew rate of the slowest comparator in the bank, and also by the decoder circuit propagation time. The decoder converts the output code to binary code needed by the computers.

The analog-to-digital converter (ADC or A/D) is a circuit that is used to produce a binary (i.e. base-2) number output that represents an analog voltage applied to the input (Figure 17-14). It is the inverse process to the digital-to-analog converter (DAC), which produces an analog output voltage that is proportional to the binary number applied to the input.

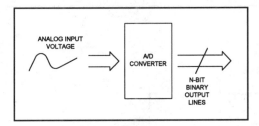

*Figure 17-14.*
*Basic function of the A/D converter is to provide a binary output that is proportional to the applied analog voltage.*

## The MAX-187 12-Bit Serial Output A/D Chip

A company called Maxim Integrated Products (120 San Gabriel Drive, Sunnyvale, CA 94086) produces a number of clever products whose design suggests their product design engineers have "been there, seen that, done it, bought the tee-shirt" and then went back home to California to design products that real engineers can use. I first became aware of Maxim products when I used some of their RS-232 interface products in a project. The product that concerns us here is one of the many A/D converter chips made by Maxim.

One reason for choosing the Maxim MAX-187 device is that readers can buy this chip for about $16 from Digi-Key (P.O. Box 677, Brooks Avenue South, Thief River Falls, MN 56701-0677; 1-800-344-4539). A lot of companies make interesting A/D converter products, but do not have distributors who are willing to sell unit quantities to hobbyists (or others who don't need a gozillion of them).

Another reason for selecting the MAX-187 is that it is available in the 8-pin miniDIP package, as well as the surface mount SO packages. Most hobbyists users will want to use the miniDIP package.

The MAX-187 (Figure 17-15) is a serial output, 12-bit A/D converter with an input voltage range of 0 to +5 volts. Because it is a 12-bit device, it will break the 5 volt input range into $2^{12} = 4,096$ different states according to the code:

| Voltage | Binary Code |
|---------|-------------|
| 0 V | $000000000000_2$ |
| 2.5 V | $100000000000_2$ |
| 4.9988 V | $111111111111_2$ |

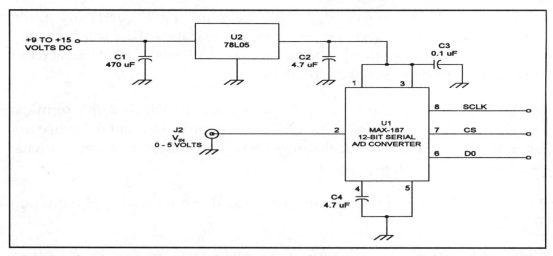

*Figure 17-15. Basic circuit for the MAX-187. The 78L05 is a +5 volt voltage regulator capable of 100 mA operation.*

Of course, value between these three points are represented with other binary numbers proportional to the actual voltage. It contains its own internal track-&-hold circuit that does not need any external holding capacitor. It also has its own internal 4.096 volt buffered reference source (i.e. $V_{REF}$ is internal to the chip). The sampling rate is 75 kHz.

The pin-outs for the 8-pin miniDIP version of the MAX-187 are detailed in Figure 17-16. These pins serve the following functions:

Pin No. 1 $V_{dd}$. DC power. This pin receives the V+ voltage, which must be between +4.75 and +5.25 volts.

Pin No. 2 AIN. Analog input. This pin receives the analog input voltage to be converted. The voltage must be kept within the range -0.3 volts and ($V_{dd}$ + 0.3) volts.

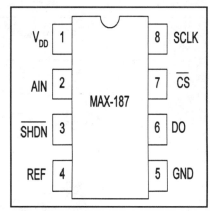

*Figure 17-16. MAX-187 pin-outs.*

Pin No. 3 SHDN (Active LOW). Shut down. This terminal can be used for providing a power saving shutdown mode. When SHDN is low, i.e. at zero volts (grounded), the MAX-187 is in the shutdown mode and consumes only 10 mW. When SHDN is tied HIGH (e.g. connected to V$_{dd}$), the MAX-187 is enabled and draws about 7.5 mW. The HIGH condition also enables the internal reference voltage. If the SHDN is left floating (i.e. no connection to either ground or V$_{dd}$), then the MAX-187 is enabled but the internal reference voltage is disabled.

Pin No. 4 REF. Reference voltage. When SHDN is HIGH, then this terminal must be bypassed to ground with a 4.7 mF capacitor (internal reference operation). When SHDN is left floating, then the external reference voltage (+2.5 to + V$_{dd}$) must be applied.

Pin No. 5 GND. Ground. Connects to the signal and power supply common line.

Pin No. 6 DO. Data out. Serial data output line. Data changes on the falling edge of SCLK.

Pin No. 7 CS (Active LOW). Chip select. Pulling this pin LOW selects the chip and initiates the conversion on its falling edge. Setting CS to HIGH deselects the chip and puts the data output line (DO) into a high impedance state.

Pin No. 8 SCLK. System clock. This clock is used to clock data out of the successive approximation register at rates up to 5 MHz.

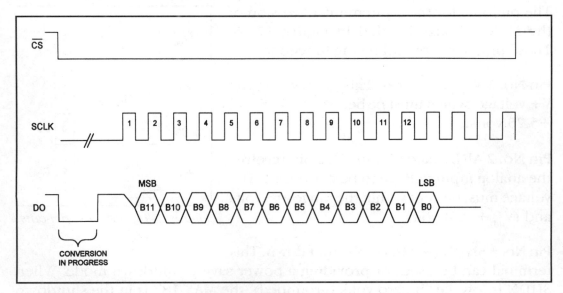

*Figure 17-17. Timing diagram of the MAX-187.*

The timing diagram for the MAX-187 is shown in Figure 17-17. The timing for the three serial interface lines (CS, SCLK, and DO) are shown. Conversion starts when the chip select line (CS, pin no. 7) is brought LOW; conversion begins on the falling edge. At this time, the data output line (DO, pin no. 6) drops LOW. The EOC is signaled by the data output line (DO) going HIGH again. At this point, clock pulses on SCLK (pin no. 8) can begin clocking data bits out through the DO line, starting with the bit B11, the MSB and finishing with bit B0, the LSB. Any interface software written for this chip must account for these factors.

One possibility for using the MAX-187 is to connect it to your computer through the parallel printer port (LPT1, LPT2, etc). Although the data lines of the printer port are outputs only, there are several handshaking lines that can be used to receive input data.

It's these lines that make all those low cost parallel printer port add-ons for PCs possible (I have an Iomega ZIP drive, an EZ-Photo scanner and several other devices that operate from my parallel printer port).

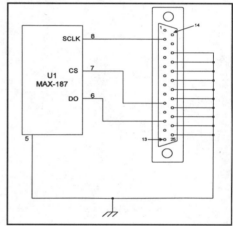

Figure 17-18. Connection of the MAX-187 to a parallel printer port interface.

The circuit for the parallel printer port interface is shown in Figure 17-18. All of the higher order lines of the DB-25 connector, from pin no. 16 to pin no. 25, are grounded. Three of the remaining pins are used for the interface:

| DB-25 Pin Number | MAX-187 Signal |
| --- | --- |
| 2 | SCLK |
| 9 | CS |
| 11 | DO |

For those who want to experiment with the MAX-187 device, Maxim makes an evaluation kit at a reasonable price. If you want to spend some more money, they also have a controller card available and accompanying software.

# Chapter 18

# Using Microcontrollers in the Instrument Front-End

A relatively new product has the potential to greatly improve the analog subsystem for sensor-based instruments, especially those that use remote sensors to a digital computer. Microcontrollers are small computers based on PIC technology, and consist of the PIC device and supporting circuitry on a small printed circuit board that can be connected into circuits as if they are a component. Two basic forms of device were reviewed when writing this chapter: the Parallax Inc. BASIC Stamp products and the MicroMint PicStic products.

Figure 18-1 shows a package outline for the Parallax BASIC Stamp model BS1-IC and the MicroMint PicStic-1, PicStic-2 and PicStic-3 devices. The 1.4 in. × 0.6 in. printed circuit board of the microcontroller mounts vertically on the printed circuit board with the other components. The connections are compatible with the 14-pin SIP socket. Examples of the actual devices, the MicroMint products, are shown in Figure 18-2. The more sophisticated Parallax BS2-IC (Figure 18-3) is a printed circuit, but is designed to mount in the socket of a 24-pin dual in-line package (DIP) integrated circuit.

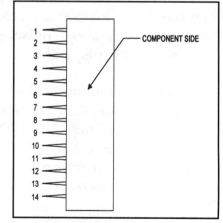

Figure 18-1. BS1-IC and PicStic-x pin-outs.

*Figure 18-2. PicStic-x products.*

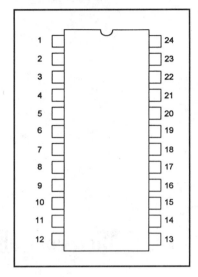

*Figure 18-3.*
*BS2-IC pinouts.*

The BS1-IC has eight bidirectional I/O lines, and executes up to 2,000 instructions per second. It will respond to up to 80 instructions. It provides a 2,400 baud serial I/O for connection to a larger computer or other instrument. Like the other products of this class, the BS1-IC will provide pulse measurements, will recognize potentiometer settings (and hence resistance sensor values), pulse width modulation, sound and other features.

The BS2-IC is a larger version of the same concept, and comes in a 24-pin DIP board. It provides 16 I/O lines, and allows up to 500 instructions which execute at a rate of 4,000 per second. The serial I/O operates up to 50 kbaud. The BS2-IC has the same features as the BS1-IC, but also offers touch-tones, frequency generation, pulse counters, serial shift registers, X-10 standard powerline controls, and other features.

There are three different PicStic products: PicStic-1, PicStic-2 and PicStic-3. In size and shape they are similar to the BS1-IC device, but add some additional features. They have 1024 bytes of EEPROM memory. The PicStic devices use the same eight I/O lines as the BS1-IC, but also add two additional I/O pins and four interrupt sources (assembly language addressable only). The serial I/O operates at rates from 300 to 9,600 baud.

PicStic-1 is very similar to the BS1-IC, except for the additional pair of I/O lines. The PicStic-2 has an on-board real-time clock, while the PicStic-3 has an on-board two-channel 0 to 5 volt, 12-bit A/D converter.

Pin-outs are:

|  | BS1-IC | PicStic-1 | PicStic-2 | PicStic-3 |
|---|---|---|---|---|
| 1 | Power | Power | Power | Power |
| 2 | Ground | Ground | Ground | Ground |
| 3 | PCO | PA3 | PA3 | PA3 |
| 4 | PCI | PA4 | PA4 | PA4 |
| 5 | +5 VDC | +5 VDC | +5 VDC | +5 VDC |
| 6 | NOT-RES | NOT-RES | NOT-RES | NOT-RES |
| 7 | PO | PBO | PBO | PBO |
| 8 | P1 | PB1 | PB1 | PB1 |
| 9 | P2 | PB2 | PB2 | PB2 |
| 10 | P3 | PB3 | PB3 | PB3 |
| 11 | P4 | PB4 | PB4 | PB4 |
| 12 | P5 | PB5 | PB5 | PB5 |
| 13 | P6 | PB6 | PB6 | PB6 |
| 14 | P7 | PB7 | PB7 | PB7 |

Where:

POWER is dc power supply 6 to 15 Vdc.

GROUND is signal and power common.

+5 Vdc is a voltage output is POWER is used, or a connection to a +5 Vdc when POWER is not used.

PCO (BS1-IC) is PC-out to computer.

PCI (BS1-IC) is PC-in from computer.

NOT-RES is an active-LOW reset.

P0-P7 (BS1-IC), PO-P15 on BS2-IC, and PB0-PB7 (PicStic) are individually configurable I/O pins.

PA3/PA4 are additional individually configurable I/O pins used on PicStic only.

The PicStic-2 requires an on-board 3-volt battery to keep the real-time clock chip active during power-off periods. The PicStic-3 has a pair of A/D converter analog inputs (ADCin0 and ADCin1) that are on the board, rather than on a pin-out.

The BS2-IC pin-outs are:

| | | | |
|---|---|---|---|
| 1 | Serial out | 13 | P8 |
| 2 | Serial in | 14 | P9 |
| 3 | RESET (HIGH) | 15 | P10 |
| 4 | Serial Ground | 16 | P11 |
| 5 | PO | 17 | P12 |
| 6 | P1 | 18 | P13 |
| 7 | P2 | 19 | P14 |
| 8 | P3 | 20 | P15 |
| 9 | P4 | 21 | +5 VDC |
| 10 | P5 | 22 | NOT-RES |
| 11 | P6 | 23 | System ground |
| 12 | P7 | 24 | POWER |

## Connecting to Serial Port

All of the microcontrollers discussed so far can be connected to the serial I/O port on a personal computer. The standard RS-232C port found on these machines normally requires ±5 to ±12 signals for HIGH and LOW, but modern machines often work quite well with standard 0 (LOW) and +2.4 to +5 (HIGH) volt signals. In those machines that will recognize TTL levels on the serial port a circuit such as Figure 18-4 can be used. In this circuit, the serial-

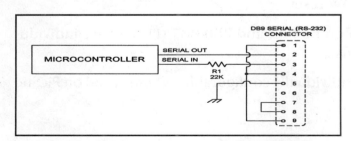

*Figure 18-4.*
*Basic connection to RS-232*
*serial port on computer.*

out pin of the microcontroller is connected to pin-2 of the DB-9 nine-pin RS-232 connector commonly used on computers today. The serial-in pin of the microcontroller is connected to pin no. 3 of the DB-9 connector through a 22 kohm resistor.

For those cases where the computer does not respond to TTL levels, or where you wish to use the standard signal levels, then use one of the RS-232C interface chips shown in Figure 18-5 (LT1181ACN) and 18-6 (MAX-232). These chips operate from +5 volts, but produce the negative signal level using a capacitor charge-pump circuit internally.

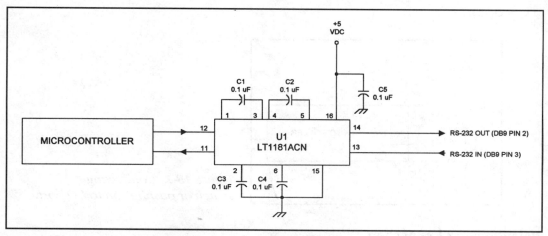

*Figure 18-5. RS-232C connection using the LT1181ACN chip.*

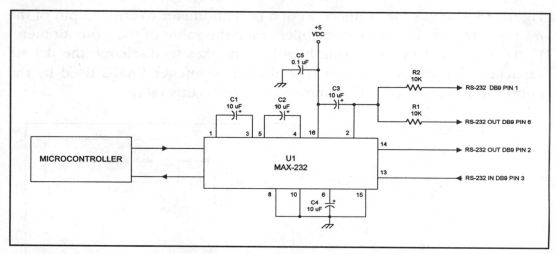

*Figure 18-6. RS-232C connection using the MAX-232 chip.*

## Pushbutton Interfacing

It is easy to interface a pushbutton or switch to the microcontrollers by using one of the I/O pins, as shown in Figure 18-7. The I/O pin is configured as an input. Resistor R1 is connected to the +5 Vdc power supply, and is used to force the input HIGH whenever S1 is open. If S1 is closed, the input pin is grounded, so is seen as LOW by the computer. The on/off state of switch S1 is read by programming the microcontroller. Some circuits omit R1, but that is not a good practice because noise can temporarily assert a LOW on the otherwise HIGH pin. The purpose of R1 is to force the pin to be HIGH when S1 is open.

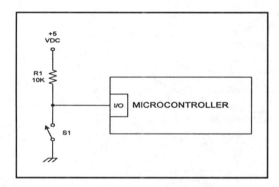

*Figure 18-7. Connecting switch or pushbutton to I/O port.*

## Potentiometer Connection

Figure 18-8 shows the connection of a potentiometer to any I/O pin of the microcontroller. The microcontroller reads the value of the potentiometer (5 to 50 kohms) by measuring how long it takes to discharge the 0.1 μF capacitor through the I/O pin. The dialect of integer BASIC used by the microcontrollers has a POT command that reads this value.

*Figure 18-8.*
*Connecting resistive sensor or*
*potentiometer to microcontroller.*

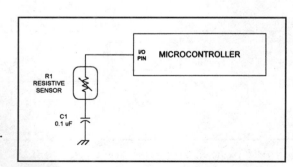

The potentiometer could be any resistive sensor (see Chapter 5), such as an actual potentiometer, a photocell, a thermistor or strain gage.

## A/D Converter

For those microcontrollers that do not have a built-in analog-to-digital converter (A/D), the circuit of Figure 18-9 can be used. This circuit uses one of several 12-bit serial output A/D converters that are on the market (LTC1298 is used here, but MAX-186 and MAX-187 can also be used with different pinouts). This particular circuit has two 12-bit, 0-5 volt A/D channels that can be selected using the software instructions programmed into the microcontroller. The Parallax manual for the BS1-IC and BS2-IC devices also shows the circuit for an ADC-0831 serial A/D converter.

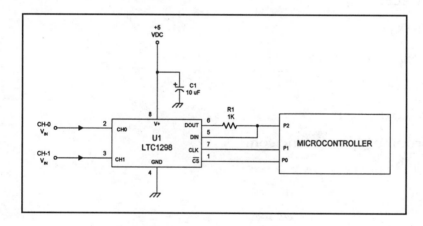

*Figure 18-9. Connecting A/D converter to microcontroller.*

## Programming

The microcontrollers can be programmed in C, assembly language, or a special version of integer BASIC called PBASIC. The manufacturers sell carrier boards for interfacing the computer (for programming the microcontroller) and other components. They also sell programming and development kits that assist the circuit developer in making the microcontrollers work in practical circuits.

## Other Sources

A number of Stamp related products can be obtained from Scott Edwards Electronics. I have used the A/D data acquisition board. Although it looks a bit pricey, it is an instrument in its own right and worth the price. Also, see Scott Edward's column in <u>Nuts 'n' Volts</u> magazine for practical projects using Basic Stamps.

Connections

Parallax Inc.

3805 Atherton Road, #102

Rocklin, CA 95765

(916) 624-8333 (voice)

(916) 624-8003 (FAX)

Web: http://www.parallaxinc.com

ftp: ftp.parallaxinc.com

e-mail: info@parallaxinc.com

List server: majordomo@parallaxinc.com


MicroMint, Inc.

4-Park Street

Vernon, CT 06066

1-800-635-3355 (voice/orders)

860-871-6170 (voice)

860-872-2204 (FAX)

860-871-1988 (BBS)

Web: http://www.micromint.com

Scott Edwards Electronics

P.O. Box 160

Sierra Vista, AZ 85636-0160

520-459-4802 (voice)

520-459-0623 (FAX)

E-mail: 72037.2612@compuserve.com

# Chapter 19

# Sensor Resolution Improvement Techniques

The role of distance, remote or indirect sensing has increased markedly over the past several decades. Such sensors are used in a wide variety of military and civilian applications in many different fields. Radar, which uses transverse electromagnetic waves to remotely detect aircraft and ships, was invented in the mid-1930s just in time for use in World War II (as the Luftwaffe found out to its sorrow during the Battle of Britain). Sonar, an underwater remote sensing scheme, uses sound waves in a manner similar to radar.

Position location sensing is used to precisely locate objects that cannot be directly observed. Metal detectors, for example, are used to find buried gas pipes, objects hidden behind walls, and other objects that are not directly viewable. Position location sensors are also used in manufacturing to locate objects moving down an assembly line, or to count objects moving along the line.

In medicine, ultrasonic sensing has been used for several decades, with rapid advances being made since the invention of the microprocessor and other digital technologies. In the 1970s, ultrasonic sensors consisted of doppler flowmeters and imaging devices that are, by current standards, primitive and crude. (Carr/Brown 1993)

## Sensor Resolution

A primary attribute of any sensing system, whether ultrasound, radar, sonar, optical or other modality, is the resolution (R) of the sensor. The definition of resolution is: "...the process of making distinquishable the individual parts of an object, closely adjacent optical images, or sources of light" (Webster, 1979). High resolution implies that closely adjacent objects can be distinguished, while low resolution systems blur the objects together.

In sensor-based imaging systems, high resolution results in sharper images in which finer grain detail is visible. More information about the structure of the imaged object can be inferred from high resolution sensors than from low resolution sensors. In most instances, higher resolution is considered superior to low resolution, although that is not universally the case.

In airborne radar and infrared sensors (e.g. "forward looking infrared" or "FLIR") a human factors issue sometimes arises regarding resolution. The high resolution sensor makes the image display too detailed for an overworked aircrew member. The best design is that which provides the user with resolution that is good enough, but not so much information as to impede rapid interpretation. In other cases, where there is time for interpretation, or where small details are of critical importance, then higher resolution is desirable.

Airborne radar resolution is usually measured on an outdoor radar range consisting of orthogonal lines of identical ground-based radar reflectors, forming a "V" shape, with spacing being close together close to the apex of the V and further apart toward the ends of each line. If the radar is flown along one line, then that line provides the range resolution while the orthogonal line provides the cross-range or azimuth resolution. The resolution is measured by counting the dots that can be distinguished from the ends toward the center, until the point is reached where they blur together forming a ragged line.

In addition to the ability to distinguish adjacent objects that are close together, there is another implication of sensor resolution: position ambiguity. When a sensor is used to precisely locate an object, the resolution of that sensor determines how accurately the position can be fixed.

Figure 19-1 shows both the position ambiguity problem as well as the source of the inability of low resolution sensors to separate two distinct objects within its field of regard. All sensors have a zone over which they detect targets, i.e. their field of regard (FOR). Targets outside of the FOR are not detected, while targets inside the FOR produce a sensor output. In the example of Figure 19-1 a generic sensor traverses past the target along a line parallel to the X-axis, producing a voltage (V) output signal. As soon as the sensor's FOR reaches the target, the sensor output begins rising. The output peaks when the target is directly opposite the sensor, which is its most sensitive point. The sensor output as the sensor location changes, and the FOR moves away from the target.

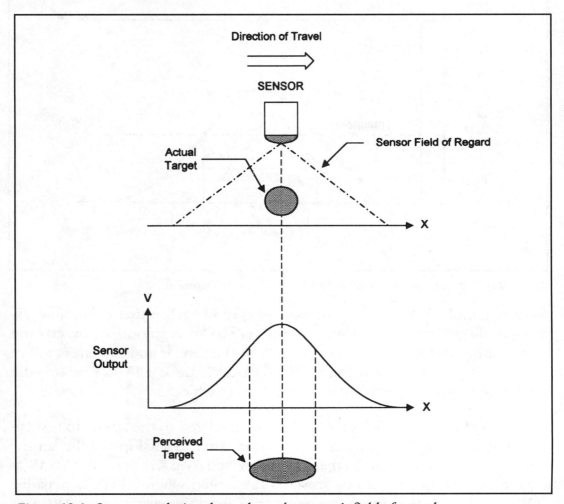

*Figure 19-1. Sensor resolution depends on the sensor's field of regard.*

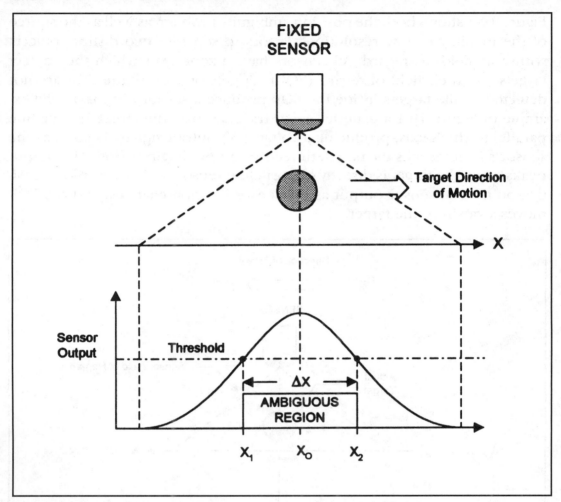

*Figure 19-2. An ambiguous zone is created by sensor resolution.*

In the example of Figure 19-1 the sensor is moving, but that is not a necessary condition. If the target moves and the sensor is stationary, exactly the same output curve results. Similarly, if both the sensor and the target are in motion, the same thing is found. The critical issue is whether or not the sensor and target are moving with respect to each other.

The ability to resolve position of the target is related to the ability to resolve differences in amplitude along the position curve. The slope of the sensor output is the discriminant in that case, i.e. when $\Delta V/\Delta X$ is positive ($\Delta V/\Delta X >$ 0), then the sensor is moving towards the target; when $\Delta V/\Delta X$ is negative ($\Delta V/\Delta X < 0$) then the sensor is moving away from the target, when $\Delta V/\Delta X =$

0, then the sensor is immediately opposite the target and the position is precisely fixed.

Or so it works in theory. The problem arises due to the inability to distinguish amplitude differences on the position curve. Several sources contribute to resolution deterioration. In analog displays, whether an oscilloscope or paper chart recording, the ability of the eye to detect small changes is often impeded.

There is also the situation where the peak of the signal is not so sharply defined as in Figure 19- 1. In those cases, the flatness of the curve creates a very small, but non-zero value of $\Delta V/\Delta X$. In digital systems, the 1-LSB step of the analog-to-digital converter, the quantization error (Q), and the noise level can adversely affect the ability to recognize small values of $\Delta V/\Delta X$, or the exact point at which $\Delta V/\Delta X \rightarrow 0$. Although the effect is shown somewhat exaggerated in Figure 19-1 for graphical reasons, the result of the problem is a smearing of the target, which distorts its shape and makes its position more difficult to fix.

The position ambiguity zone can sometimes be adjusted by a threshold arrangement similar to Figure 19-2. Threshold detector systems are sometimes seen in which an adjustment is permitted to remove false targets, noise components and other artifacts. The threshold is adjusted to be above the noise and error floor, but this does not affect the inability to detect the point at which $\Delta V/\Delta X$ goes to zero.

A superior approach to reducing the ambiguity zone is to increase the resolution of the sensor, if it is possible with the specific sensor in question. This task is done by selecting a sensor with a smaller field of regard (Figure 19-3). The output voltage-vs-position (V vs X) is therefore narrower, and the maximum extent of the ambiguous region is correspondingly smaller. The distortion of the target shape, as well as the position ambiguity is improved. Again, this approach does not address the problem of detecting the $\Delta V/\Delta X = 0$ condition.

It must also be noted that reducing the FOR, as in Figure 19-3, may not be either possible or feasible in all cases. If the instrument already uses a sensor with the highest available resolution, but is nonetheless still insufficient, then the problem is not solved.

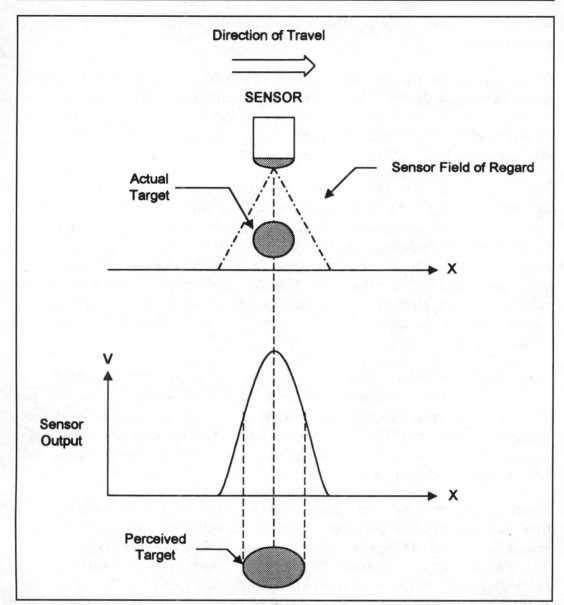

*Figure 19-3. Narrowing the resolution reduces the ambiguity zone.*

The target discrimination problem is shown in Figure 19-4. In this scenario, a pair of identical targets, T1 and T2, are spaced a distance ($\Delta X$) apart, with T1 centered on X1 and T2 centered on X2 (Figure 19-4a). The mid-point between T1 and T2 is designated $X_o$. A sensor traveling past the targets, on a line parallel to the line of centers T1-T2, will begin producing an output as soon as the FOR encounters T1. This signal is shown as V1. The signal V1 will peak when the sensor is opposite the target, i.e. at X1.

When T2 enters the sensor's FOR, it begins to contribute to the sensor output. The contribution of the second target is shown as V2. The actual output of the sensor is a composite of V1 and V2, as shown in Figure 19-4b. The characteristic double-hump results from the difference in position between

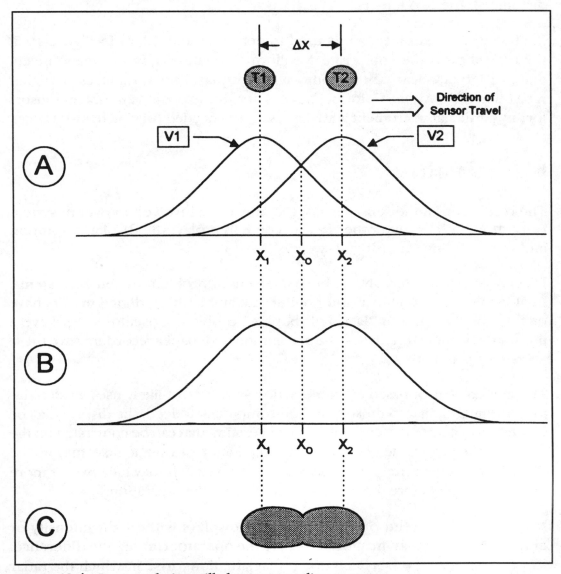

*Figure 19-4. Poor resolution will obscure two adjacent targets.*

the peaks of V1 and V2. The graphical representation of the two targets is similar to Figure 19-4c. The peals are more or less distinguishable, but the two segments are connected together. The "dumbbell" shape makes it difficult to discern the boundaries between T1 and T2. When the resolution is poorer yet, then the dumbbell deteriorates into an oblong shape similar to Figure 19-1. In that instance, the dip seen in the signal curve of Figure 19-4b becomes more shallow, or disappears altogether. It is then impossible to distinguish the two targets.

If a higher resolution sensor is used to detect T1 and T2, as in Figure 19-5, then the dip on the composite waveform becomes far more pronounced, and the targets clearly break out on the display. The advantages of higher resolution are obvious. Unfortunately, it's not always possible to obtain sensors for any particular instrument with the resolution needed (of which, more later).

## Other Resolution Issues

The resolution of the sensor is not the only issue involved in overall system resolution. At least two other factors come into play: the display resolution and the sampling rate (digital systems).

Display resolution has always been a potential problem in sensor systems. Analog meters have inertia and pointer size ambiguities; digital meters have quantization error and "last digit bobble" problems; oscilloscopes have a minimum beam width that can be accommodated; paper recorders have these and other problems.

The modern sensor-based instrumentation system most likely uses a video display terminal similar to those found on computers. Indeed, the display may be identical to a computer terminal. The smallest unit that can be controlled on the display is the picture element, or pixel. A separation of a single pixel may not be discernible on the screen if the pixel size is too small. It may take two or more pixel separation before the user's eye can discern the separation.

Another problem seen on digital or video displays is the light intensity or gray scale change that must occur before the operator can see the difference. In work on radars, it has been noted that situations arise in which the radar

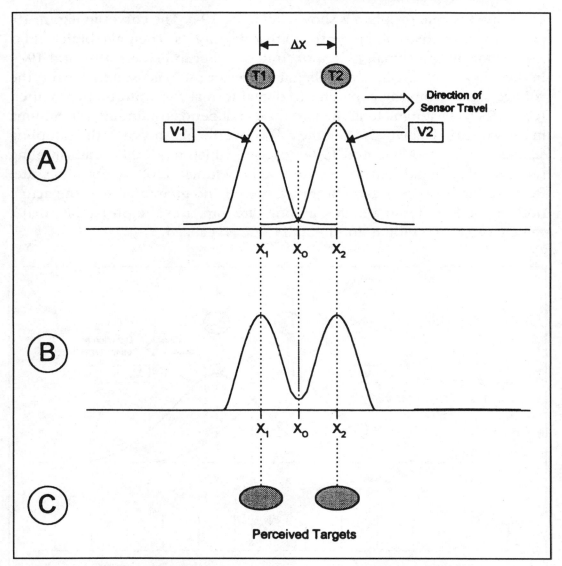

*Figure 19-5. Improving resolution makes it possible to breakout both targets.*

signal processor can distinguish two targets, and the data stream from the processor proves it, yet the operator cannot. In some cases, there was a separation of several pixels, but the problem turned out to be that the intensity of the light on the screen only reduced 1-LSB 2 × 1-LSB. There may have to be a minimum specified change in signal intensity for the operator to see the difference, even though examination of the data stream would clearly show the difference.

The sampling rate problem is shown in Figure 19-6. The curves in Figure 19-6a are the responses of the sensor to the two targets. Their algebraic sum is the sensor output signal, as demonstrated earlier in Figures 19-4 and 19-5. In digital systems, an analog-to-digital converter (A/D) is used to convert the voltage signal from the converter to digital format for input to a computer. (Carr 1992) Distinguishing the two targets depends on finding the minima in the value of data words from the A/D converter. As shown in the sampling signals of Figures 19-6b and 19-6c (each of which has a different sampling rate), the depth and sharpness of the null is a function of the sampling rate. Because the process is asynchronous, there is no guarantee that the actual null will be found, but a high sampling rate improves the probability that a value close to the null, if not the null itself, is realized.

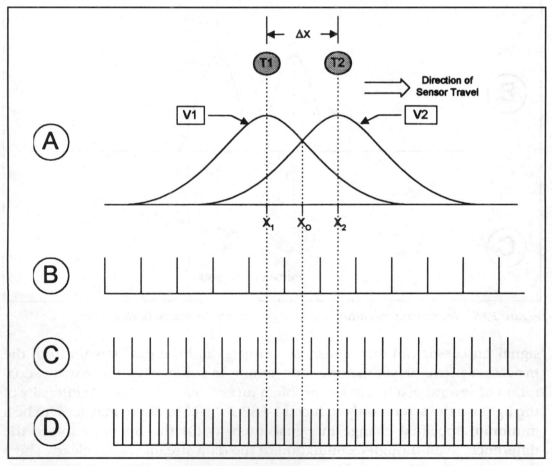

*Figure 19-6.  Sampling rate problems can also cause resolution problems.*

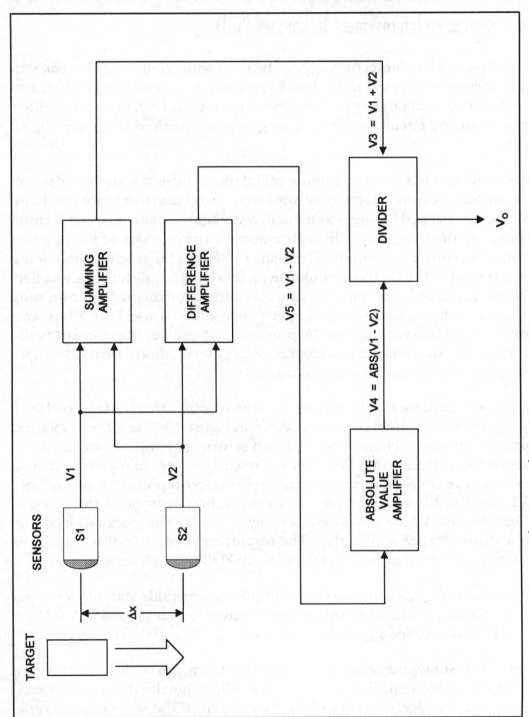

*Figure 19-7. Block diagram of sensor resolution improvement system.*

## Sensor Resolution Improvement Techniques (SRI)

A technique from radar technology can be used with other forms of sensor to greatly increase the resolution. Two forms exist. One uses two sensors, and the other uses a single sensor. The two-sensor method can be used in either digital or analog circuits, where the single-sensor method is used in digital circuits.

Radar systems often use a technique called monopulse resolution improvement, or MRI. (Skolnik 1980, Stimson 1983) Two basic forms are used. Another method, called lobing, sequentially switches the main lobe (or "central maxima") of the beam from the radar antenna from one side of the target to the other. Comparisons can then be made to discern the true location of the target. (Stimson 1983) The monopulse methods accomplish the same effect by breaking the antenna into two sections, and simultaneously transmitting two different lobes towards the target (some systems use four lobes, two horizontal and two vertical, but the principle is the same). By comparing the return signals, the radar can locate the target. Two methods are used: amplitude comparison and phase comparison.

Amplitude comparison can be used for sensors other than radar. The block diagram for a two-sensor version is shown in Figure 19-7, and the associated signals are shown in Figure 19-8. The two sensors (S1 and S2) are aligned a distance $\Delta X$ apart such that their fields of regard overlap. In my experiments, the FORs were overlapped at approximately the -3 dB points (-6 dB voltage) of V1 and V2. As the target traverses along a line in front of the sensors, voltages V1 and V2 are produced, see Figure 19-8a. These signals have the characteristic shapes seen earlier. The resolution of these individual signals are determined by the size and shape of the FOR for each sensor.

The processing for sensor resolution improvement (SRI) starts by combining the signals V1 and V2 in two different ways: the sum signal (V3 = V1 + V2) and the difference signal (V5 = V1 - V2).

Figure 19-8b shows the sum signal. Note that the width of this signal is considerably broader than that of either individual signal, which is expected because it is the algebraic sum of the two signals. If the sensors were separated more, then a dip would appear in the center of the waveform.

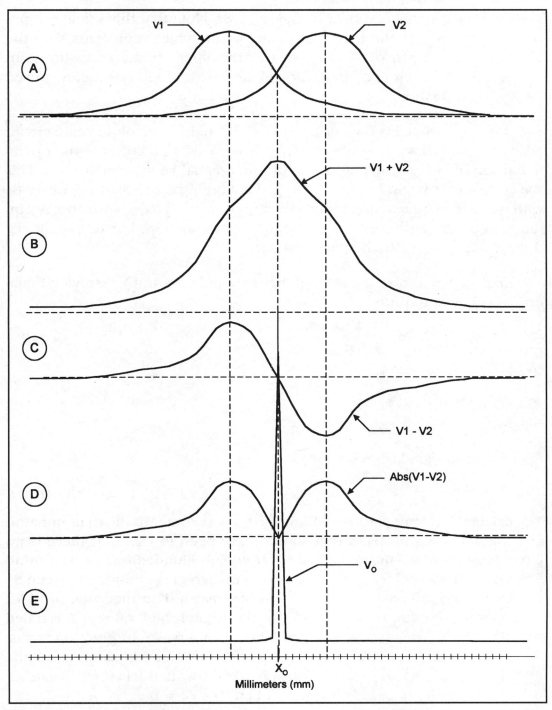

*Figure 19-8. Sensor resolution improvement curves: a) V1 and V2; b) V1 + V2; c) V1 - V2; d) Abs (V1 - V2); e) $V_o$.*

The difference signal is shown in Figure 19-8c. Note that this signal is bipolar. It is positive on one side of the center point, which represents $X_o$ in the $\Delta X$ space between sensors, and negative on the other. The zero-crossing point represents a relatively good indication of the position, and is sometimes used for $X_o$ discrimination.

The difference signal is further processed through an absolute value circuit (also called a full-wave rectifier or full-wave precision rectifier in the operational amplifier implementation) to produce signal V4 = Abs (V1 - V2). This waveform is shown in Figure 19-8d, and is nothing more than Figure 19-8c with the negative excursion flipped over to make both halves positive. Again, the zero point can be used for position discrimination, but does little for distinguishing two closely spaced targets.

The final step is to take the ratio of the two signal, V3 and V4, solving Equation 19-1 below to find $V_o$:

$$V_O = \frac{V1 + V2}{k + (V1 - V2)}$$

eq. (19-1)

Where:

$V_o$ is the output signal.

V1 is the signal from sensor S1.

V2 is the signal from sensor S2.

k is a small constant.

The original implementation did not use the k constant in the denominator, and that caused a problem that should have been foreseen. If there is no target present, which means S1 and S2 are equally illuminated by background light (optical sensors were used), or when the target is mid-way between S1 and S2 (which can occur with any form of sensor), then the value of V1-V2 goes to zero. This condition forces $V_o$ to an extremely high value that is called "infinity" in the naive sense and "undefined" otherwise. In practical terms, this means that the amplifier or divider at the output will saturate, or the digital value tries to go to above full-scale. The constant k is set to whatever value will force $V_o$ to exactly full-scale when V1 - V2 = 0.

The resolution improved output $V_o$ is shown in Figure 19- 8e. Notice the tremendous improvement in resolution compared with the beam widths of V1 and V2, as shown in Figure 19-8a. The implication of this change to both position ambiguity and two-target discrimination is obvious. This waveform is an extremely high amplitude, narrow base, spike that accurately finds $X_o$.

In Figure 19-7, the block diagram assumes analog implementation, so the stages are labelled "amplifiers." The sum amplifier, the difference amplifier and the absolute value amplifier were operational amplifiers. The divider can be any of several analog multiplier/divider integrated circuits that are on the market, although I used the Burr-Brown DIV-100 module. The same block diagram can be used as the basis for a digital or computer implementation by programming the stages using math statements. Even ordinary BASIC can be used for this purpose.

## Simulation Effort

The first step in demonstrating this principle involved a computer simulation on artificial data using an Excel 5.0 spreadsheet. A normal distribution ("bell-shaped") curve drawing template was used to draw curves representing V1 and V2 on fine-mesh linear graph paper with 1 mm (»25/inch) spacing between lines.

Approximately 80 values for V1 and V2 were obtained by counting the amplitudes of the two curves at 1 mm intervals along the baseline. These values were entered into the "A" and "B" columns of the spreadsheet. Another column was programmed with a math function to solve Equation 19-1, and insert the result in the cell adjacent to the V1 and V2 cells from which the value was calculated. The chart function of Excel was then used to show V1, V2, V1+V2, V1-V2, ABS(V1-V2) and $V_o$. The result was consistent with the expectations of Figure 19-8. With that encouragement a working model was produced.

## Laboratory Model of the Two-Sensor Approach to SRI

The original goal was to build a high-resolution magnetometer based on the Speake & Co. Ltd. FGM-3 flux-gate sensor. This sensor measures the local magnetic field, and then converts it to a proportional frequency. A set of special purpose "application specific integrated circuits" (ASIC) produced by Speake & Co. Ltd converts the frequency to an 8-bit binary number (several ASICs are available including a magnetometer, a gradiometer and a compass). The 8-bit parallel outputs of the ASICs can be either fed directly to a computer, or to an digital-to-analog converter (DAC) that will produce a voltage or current output proportional to the strength of the magnetic field.

Experimenting with magnetic sensors can be tricky in environments where the local fields are difficult to either predict or control. In order to simplify the test, an optical sensor pair was used (Figure 19- 9).

The optical sensors were Burr-Brown OPT-101 devices. These sensors are operational amplifiers with an photodiode device built-in to the transparent 8-pin DIP IC package. The two OPT-101 devices were spaced 10 mm apart so that their cones of acceptance overlapped (which is also the minimum possible X-axis separation due to the size of the IC packages). The mid-point between the two OPT-101 devices corresponds to $X_o$, while the distance between corresponds to DX; any particular point along the path between S1 and S2 is designated $X_i$.

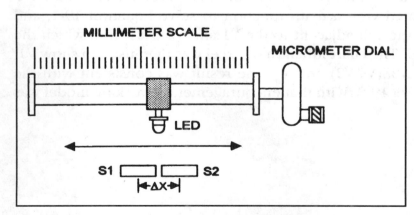

*Figure 19-9. Test set-up using optical sensors to test concept.*

The target was a red light emitting diode (LED) mounted on a micrometer device that measured distances in millimeters. The initial position of the LED target was set so that it was outside the FOR of both S1 and S2. The LED was then advanced 1 mm at a time until it had traversed the entire distance from the left-most extent of the FOR of S1 to the right-most extent of the FOR of S2. The output voltages (V1 and V2) from S1 and S2 were measured with a 3-1/2 digit digital voltmeter (DVM) at each 1-mm interval. Those data were then entered into an Excel spreadsheet and plotted on a chart. The curves in Figure 19-8 were cut-and-pasted directly from the Excel chart to a Visio Technical 4.0 drawing, so represent actual results.

## Single-Sensor Method

The two-sensor method of SRI produces startling results, but at the cost of two sensors. It is easy to implement in analog circuitry, and can also be implemented in digital circuitry. A prototype ASIC device (JC101R) that digitally implements the two-sensor method was supplied by Speake & Co. Ltd for evaluation. Another approach uses a single sensor and a look-ahead calculation to synthesize V2. It is easily implemented digitally, but is somewhat difficult to implement in analog circuitry.

Assume that the values of V1 from the sensor are a series $V_i$ in which each value represents the signal amplitude at sequential locations along the X-axis:

$$V1\ V2\ V3\ V4\ V5\ V6\ V7\ V8\ V9\ V10\ V11\ ...\ V_{i+N}$$

Each value of $V_i$ represents a value of V1 in Equation 19-1 above. The corresponding value of V2 in Equation 19-1 is found by taking a subsequent value of V1 that is displaced a distance N (an integer) from $V_i$. Thus, in terms of Equation 19-1:

$$V1 = V_i$$
$$V2 = V_{i+N}$$

Equation 19-1 can be rewritten to the form:

$$V_o = \frac{V_i + V_{i+N}}{k + (V_i - V_{i+N})}$$

eq. (19-2)

When the V1 data from the experiment are plotted, the resultant curves are very similar to those of Figure 19-8 and demonstrate very nearly the same degree of SRI. It was also found that a limited amount of "tuning" of resolution can be done by selecting values of N. However, with the 1 mm spacing used in the experiment, values N > 5 showed essentially the same curves.

In both the single-sensor and two-sensor methods there exists the possibility of creating a selectable beam width sensor system. Signal $V_o$ could be the narrow beam width signal, V1 + V2 can be the wide beam width signal, and V1 (or V2) can be the medium beam width signal.

## Conclusion

The sensor resolution improvement (SRI) method is a variant on the monopulse resolution improvement (MRI) method used for many years in radar technology. It appears to have application in ultrasonic imaging (which resembles radar in basic approach), and any instrumentation problem where sensor resolution is an issue.

# Chapter 20

# DC Power Supplies for Sensor-Based Circuits

The dc power supply is critical to the success of any electronic circuit, perhaps especially instrumentation circuits. This chapter deals mostly with the control and distribution of dc power, with some special emphasis on reference supplies and excitation power supplies needed for resistive and other sensors. We will look at the standard dc power supply circuit only in passing for those who may not be familiar with it.

## DC Power Supplies

The purpose of the dc power supply is to convert alternating current (ac) to direct current (dc) for use in electronic circuits. Figure 20-1 shows the basic circuit diagram for the dc power supply. Transformer T1 scales the voltage from 120 or 240 Vac to the voltage range needed for the electronic circuits (typically 5 to 28 Vdc).

The ac from the transformer is converted to dc in the bridge rectifier BR1. These four-terminal devices take the ac voltage at the terminals marked "ac" and outputs them through the dc terminals ("+" and "-"). The output of the rectifier is pulsating dc with a peak voltage about 0.9 times the peak ac voltage. The voltage rating for the transformer is in RMS voltage, so the output of the rectifier is $0.9 \times 1.414 \times V_{RMS}$, or $1.27V_{RMS}$, where $V_{RMS}$ is the voltage appearing across the secondary winding of the transformer.

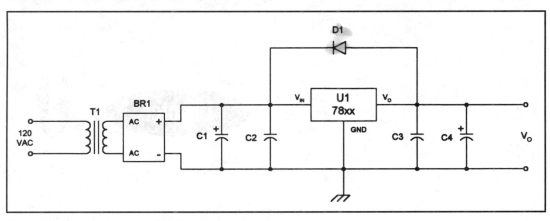

*Figure 20-1. Typical regulated voltage supply.*

The voltage regulator (U1) keeps the output voltage constant despite variations of input voltage or variations of load current. There is usually a 2.5 to 3 volt drop between the $V_{in}$ and $V_o$ terminals of the voltage regulator. Thus, the transformer secondary RMS voltage must be $\geq (V_o + 3)/1.27$.

The voltage regulator is usually a three-terminal IC device of either LM-340n-xx or 78xx types (essentially the same thing under different nomenclatures). In LM-340 devices, the "n" is replaced with the package style indicator: L for TO-92, K for TO-3, and T for TO-220. The "xx" is the voltage rating. Thus, an LM-340T-5 is a +5 volt regulator in a plastic TO-220 package. For negative output regulators, the LM-320n-xx series are used. In 78xx devices, the "xx" is the voltage rating, so a +5 volt type is 7805, while a +12 volt is 7812. If the device is rated at 100 mA or so, then an "L" is placed between the 78 and xx: 78L05 is a 100 mA, +5 Vdc device. Negative output equivalents are designated 79xx.

The current ratings for most regulators are:

| | |
|---|---|
| TO-92 | 100 mA |
| TO-220 | 750 mA (1,000 mA with heatsink) |
| TO-3 | 1,000 mA (some are 3,000 mA) |

Most common devices are either TO-92 (e.g. 78L05), or TO-220 (e.g. 7805), producing 100 mA and 750 mA respectively.

Capacitor C1 in Figure 20-1 is rated at 2,000 µF per ampere, except for low current devices in which case 1,000 µF is used; for the standard 750-1000 mA supply a 2,000 µF unit is used. The exact rating is not important, so long as the right range is achieved. The output capacitor (C4) is optional, and is used to smooth variations occuring before the regulator can catch up, i.e. transients. This capacitor is rated nominally at 100 µF per ampere. In both C1 and C4, the voltage rating of the capacitor should be at least 150 percent of the applied voltage rating, more if possible.

Capacitors C2 and C3 are used for noise suppression. These capacitors are needed to take care of transients coming across the ac power supply line. Make these capacitors 0.1 µF to 1 µF.

Diode D1 is used only if C4 is used. This diode should be a 1N4007 device. It is used to rapidly discharge C4 when power is turned off, preventing it from biasing the substrate of U1, causing its destruction.

## DC Power Distribution

The dc power supply is one of the most important, but least considered, aspects of electronic project design. The elements of dc power supply design are well known, and have been covered in depth in these pages. But what is often missing is how to distribute the power throughout a project. And that little matter is of critical importance. When I was in engineering grad school, I worked for the university repairing medical equipment. A fairly large number of grad students and senior undergrads doing various projects found their way to my workshop with one problem or another. Very frequently, the design of the circuit was fine, but the implementation was a real problem. In most cases, they created their own bad situation by paying too little attention to the power supply distribution. The main errors were too small of conductors and no decoupling capacitors. One fellow came into my shop with a perf-board covered with about thirty TTL chips, and complained that it worked only some of the time. After ascertaining that the problem was probably not a loose soldering connection or broken wire, I noticed that the wire to each chip was very small (#30 wire-wrap wire), and that there were no bypass capacitors anywhere on the perf-board. I instructed him to put

470 μF at the +5 Vdc input line, plus 0.1 μF at each TTL chip...and increase the gauge of the wire distributing power. When he made those changes, the project worked properly.

In the sections below you will find out why these are important.

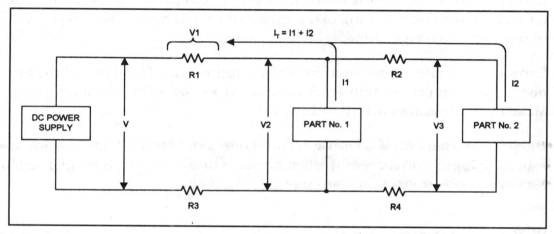

*Figure 20-2. The problem when two or more circuits share a dc power supply.*

Figure 20-2 shows a generic electronic circuit dc power distribution system. A dc power supply, which may or may not be voltage regulated depending on the application, feeds dc to to the individual circuits via a common dc power line and a ground. In an ideal world, that would be the end of our concerns. But in real circuits, there is a dc resistance associated with all of the power line conductors and any connectors used in the system. These resistances are shown as R1 through R4 in Figure 20-2. As current drawn by each part flows through the conductors, a voltage drop will appear across the conductor resistance. For example, voltage V1 is the IR drop across R1 due to the total current in the circuit.

Two things happen because of those voltage drops. First, the actual voltage available to each part is lower than the supply voltage by the amount of the voltage drops. For example, Part-1 can see a maximum voltage (V2) of V-V1. When selecting a value for V you must either take into consideration the voltage drops that will occur, or reduce the resistance to a negligible value by making the conductors large enough.

The second thing that can happen is that varying currents in one part will cause the supply voltage to vary at that part and at other parts. For example, if the dc load represented by current I1 varies with the signal in Part-1, then the total current varies, and that means the voltage drop across R1 also varies. That varying dc supply voltage is applied to Part-2 as its supply voltage. Signal can thus be coupled unintentionally from one stage to another through the dc power supply lines. This effect is usually called power supply sensitivity (PSS), and certain components (e.g. operational amplifiers) have a PSS specification.

The power supply dc return lines, which are usually also the grounds for the circuit, have the same kind of resistances as the "hot" line. If the same conductors are also used as signal grounds (often the case), then variations in ground voltage drops are seen as valid signals between stages; these are called ground loops. The equivalent circuit would show the varying dc from the ground loop in series with the normal signal applied to the stage. In a multistage circuit, the combination of voltage drops and spurious signal paths can be quite complex, and very difficult to troubleshoot.

Fortunately, there are some things that you can do about it. One thing, of course, is to be generous with the wire size used in the circuit for dc power lines and grounds. I've seen problems caused by using #24 hook-up wire where #18 would've been more appropriate. Also, I've seen cases where printed wiring tracks were too narrow and caused excessive voltage drops in relatively short lengths. My first homebrew microcomputer had that problem. Because it contained a large number of old-fashioned TTL devices, each drawing 25 mA, the current requirement was huge.

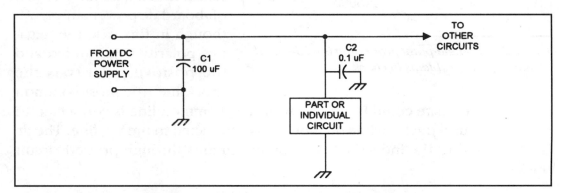

*Figure 20-3. A decoupling capacitor (C1 & C2) at each stage will work wonders.*

Another thing that you can do about the problem is to be generous with the number of capacitors used in the circuit. The capacitors (e.g. C1 and C2 in Figure 20-3) serve two functions: they decouple noise signals on the power line (prevent them from being coupled from stage to stage), and they serve as local resevoirs of current to fill in for short-duration current needs of the circuit being served. It has long been the practice in digital circuits to put a 0.1 mF capacitor across the power terminals of each chip on the circuit board. Putting a 1 µF or higher capacitor at each analog circuit also works well.

Another thing that can be done is to use "single-point grounding" techniques. Each circuit should have its own ground system, and all of the individual ground systems are connected to the chassis at only one point. Also, if you mix different kinds of grounds in the same project, then each type should have its own ground bus. In the example shown in Figure 20-4, there are separate grounds for dc power, analog signals and digital signals.

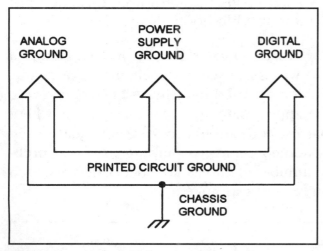

Figure 20-4. Single-point grounding is well-suited to keeping the problems in check.

## Power Distribution Schemes for Circuit Boards

There are a number of different approaches to managing dc power on projects with printed circuit boards. The version shown in Figure 20-5 is the simple case of an individual circuit card served by an off-board dc power supply. Although in this case the negative polarity is grounded and the positive polarity is the "hot" line (the most common case), the opposite could be true as well. The common line is connected to the PCB ground tracks, while the hot line is connected to the V+ line. The dc is distributed to the individual parts or sub-circuits through printed circuit tracks.

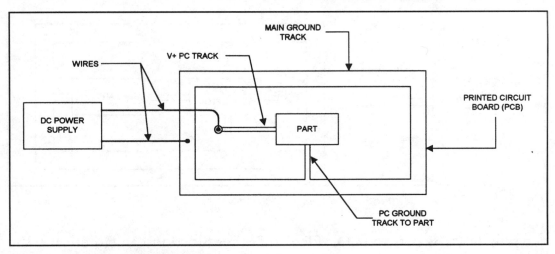

*Figure 20-5. Simple dc distribution system.*

A different, but related, scheme is shown in Figure 20-6. Here there may be more than one individual printed circuit board with individual circuits, and all are plugged into a motherboard. The motherboard may be solely for distribution of signals and dc power (as in the old S-100 microcomputers of the late 1970s and early 1980s), or there may be some additional circuitry on the motherboard. On the IBM-PC compatible computers, for example, the motherboard contains the processor and main functions of the board, and

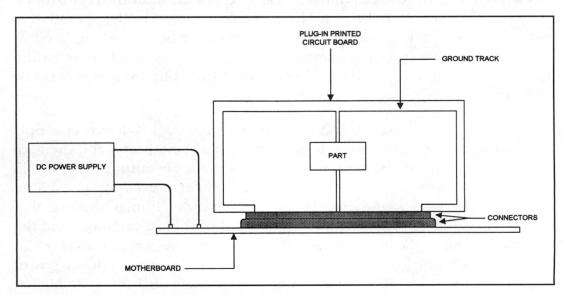

*Figure 20-6. Use of a motherboard and plug-in PCBs.*

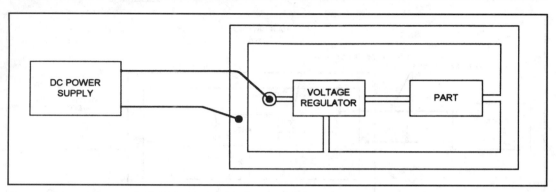

*Figure 20-7. The use of an on-board voltage regulator.*

the plug-in PCBs perform I/O, video and other functions. In either case, the dc power is distributed to the various printed circuit boards via the motherboard and the plug-in connectors that mate the motherboards to the plug-in boards.

The situation in Figure 20-6 assumes that a single dc power supply serves the needs of all the PCBs plugged into the motherboard. In some cases, however, if current drains are high or there is a lot of digital (or other) noise generated on the PCBs, then one might wish to use an approach like Figure 20-7. This method is called distributed voltage regulation because each plug-in card has its own voltage regulator. The effect of the multiple regulators is to isolate the circuits on the board from ill effects caused by the operation of the other boards. The main dc power supply may be either regulated or unregulated. If a regulated main power supply is used, it is called preregulation, and is capable of a very good degree of power supply ripple reduction.

The power supply sense method is shown in Figure 20-8. A dc power supply with sense lines samples the voltage present at a remote site, and then adjusts its own voltage set-point to put the power supply output voltage high enough to meet the requirement at the remote site. For example, in a microcomputer that uses a motherboard and plug-in cards, it might be true that the processor card (PCB-2) is the most critical of voltage variation, and the memory cards are the least critical (PCB-1 or PCB-3). As a result, the builder will place the sense lines closest to the processor card, causing the dc power supply voltage to be +5 Vdc at that point. Closer to PCB-1 it will be higher, and towards PCB-3 it will be lower.

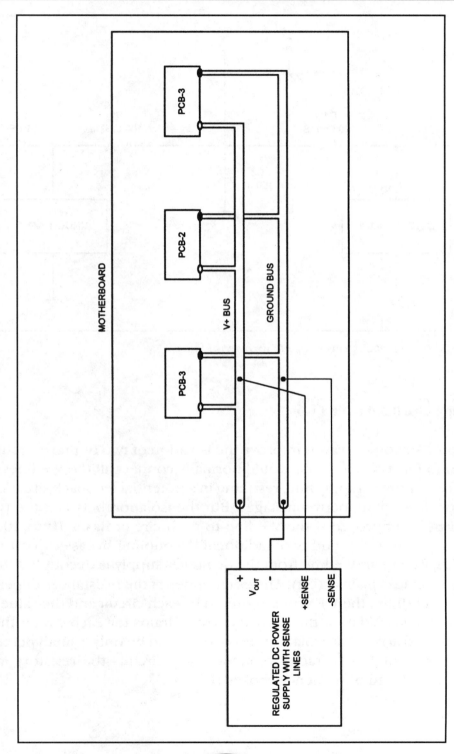

*Figure 20-9. Isolation resistors or RF chokes will often improve isolation between stages.*

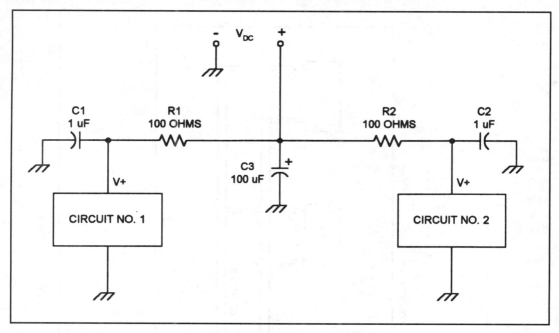

*Figure 20-8. Use of dc power supply with sense lines.*

## Isolating Circuits on The PCB

Figure 20-9 shows a way to improve the isolation of two or more circuits in a group. In most circuits, one would normally connect all the V+ lines to the positive dc power supply line, resulting in a potential feedback situation due to poor decoupling between stages. But the isolation between the two circuits can be improved if the V+ line to each circuit has a 10 to 100 ohm resistor (R1 and R2), and two additional decoupling bypass capacitors (C1 and C2). The main V+ line from the dc power supply is decoupled through a higher value capacitor (C3). The exact values of the resistances depends on the current drain, the V+ values required by each circuit and the value of the supply voltage (Vdc). At the normal current drains for each circuit the voltage dropped across the isolation resistors should be only a small percentage of the total voltage. In radio frequency (RF) circuits the resistors may be replaced with radio frequency chokes (RFCs).

# Works Cited

American Society for Testing and Materials. *Manual on the Use of Thermo-couples in Temperature Measurement*.

Carr, Joseph J. and John M. Brown. *Introduction to Biomedical Equipment Technology - 2nd Edition*.

Carr, Joseph J. *The Art of Science*.

Deming W. Edwards. *Out of the Crisis*.

Herceg E. E. *Handbook of Measurement and Control*.

Janicke, J.M. *The Magnetic Measurements Handbook*.

Klier, George. "Signal Conditioners: A Brief Outline," <u>Sensors: The Journal of Machine Perception</u>.

Noble, Richard. "Fluxgate Magnetometry," <u>Electronics World and Wireless World</u>.

Sheingold, Daniel H. (ed.). *Transducer Interfacing Handbook*.

Skolnik, Merrill *Introduction to Radar Systems*.

Stimson, George W. *Introduction to Airborne Radar*.

*Webster's New Collegiate Dictionary*.

**Notes**

[1] NIST is National Institute for Standards and Technology, formerly called National Bureau of Standards.

[2] Definitions derived in part from George Klier. Some of the definitions used in this article were organized in the Klier article previously cited. For additional information see the Daniel H. Sheingold reference.

[3] -6 db is used when voltage is measured; -3 db is used when power is measured.

# Index

## A

## C

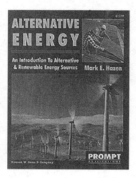

## Semiconductor Cross Reference Book
## Fourth Edition
*Howard W. Sams & Company*

This newly revised and updated reference book is the most comprehensive guide to replacement data available for engineers, technicians, and those who work with semiconductors. With more than 490,000 part numbers, type numbers, and other identifying numbers listed, technicians will have no problem locating the replacement or substitution information needed. There is not another book on the market that can rival the breadth and reliability of information available in the fourth edition of the *Semiconductor Cross Reference Book*.

**Professional Reference**
688 pages ✦ Paperback ✦ 8-1/2 x 11"
ISBN: 0-7906-1080-9 ✦ Sams: 61080
$24.95 ($33.95 Canada) ✦ August 1996

## IC Cross Reference Book
## Second Edition
*Howard W. Sams & Company*

The engineering staff of Howard W. Sams & Company assembled the *IC Cross Reference Book* to help readers find replacements or substitutions for more than 35,000 ICs and modules. It is an easy-to-use cross reference guide and includes part numbers for the United States, Europe, and the Far East. This reference book was compiled from manufacturers' data and from the analysis of consumer electronics devices for PHOTOFACT® service data, which has been relied upon since 1946 by service technicians worldwide.

**Professional Reference**
192 pages ✦ Paperback ✦ 8-1/2 x 11"
ISBN: 0-7906-1096-5 ✦ Sams: 61096
$19.95 ($26.99 Canada) ✦ November 1996

# CALL 1-800-428-7267 TODAY FOR THE NAME OF
# YOUR NEAREST PROMPT PUBLICATIONS DISTRIBUTOR

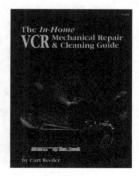